OCEANS

OCEANS

A VISUAL GUIDE

Stephen Hutchinson
Lawrence E. Hawkins

FIREFLY BOOKS

A FIREFLY BOOK

Published by Firefly Books Ltd. 2008

Conceived and produced by Weldon Owen Pty. Ltd.
59 Victoria Street, McMahons Point, Sydney, NSW 2060, Australia

First printing

Publisher Cataloging-in-Publication Data (U.S.)
Hutchinson, Stephen.
  Oceans : a visual guide / Stephen Hutchinson and Lawrence E. Hawkins.
Originally published: London : The Reader's Digest Association Limited, 2004.
[304] p. : col. ill., col. photos. ; cm.
Includes index.
Summary: A guide to oceans, their formation and extent, life-forms they
support, their value to humans, and the threats they face.
ISBN-13: 978-1-55407-069-5 (bound)   ISBN-13: 978-1-55407-427-3 (pbk.)
ISBN-10: 1-55407-069-4 (bound)   ISBN-10: 1-55407-427-4 (pbk.)
1. Ocean. 2. Oceanography — Miscellanea. I. Hawkins, Lawrence E. II. Title.
551.46 dc22   GC21.H8834   2008

Library and Archives Canada Cataloguing in Publication
Hutchinson, Stephen
  Oceans : a visual guide / Stephen Hutchinson, Lawrence E. Hawkins.
Includes index.
ISBN-13: 978-1-55407-069-5 (bound)   ISBN-13: 978-1-55407-427-3 (pbk.)
ISBN-10: 1-55407-069-4 (bound)   ISBN-10: 1-55407-427-4 (pbk.)
1. Ocean. 2. Oceanography. I. Hawkins, Lawrence E. II. Title.
GC21.H885 2005   551.46   C2004-907208-0

Published in the United States by
Firefly Books (U.S.) Inc.
P.O. Box 1338, Ellicott Station
Buffalo, New York 14205

Published in Canada by
Firefly Books Ltd.
66 Leek Crescent
Richmond Hill, Ontario L4B 1H1

Illustrators: Peter Bull Art Studio; Tom Connell/Wildlife Art Ltd.;
  Moonrunner Design
Designers: Hilda Mendham; Jo Raynsford
Cover Design: Jacqueline Hope Raynor

Cover photo credits
Front cover and spine: Tim McKenna/Corbis
Back cover (left to right)**:** Photolibrary.com; Australian Picture Library/
  Corbis; Seapics.com/Mark Conlin
Front flap: Australian Picture Library/Corbis

Color reproduction by Chroma Graphics (Overseas) Pte. Ltd.
Printed in China

Contents

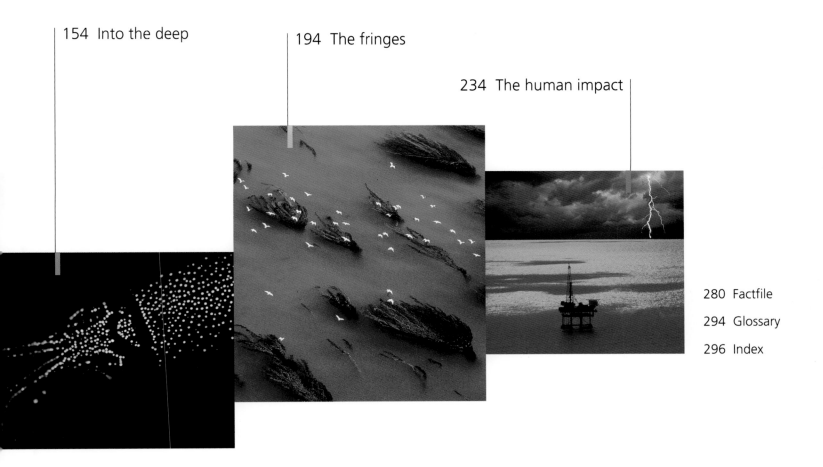

Introduction

A famous oceanographer used to start his lectures on the oceans with the phrase, "The oceans are wide, deep, dark, cold and salty." He could have added that, although it constitutes 71 percent of Earth's surface, much of the vast ocean realm remains to be explored. The presence of liquid water on the surface of our planet is remarkable in itself, and it is Earth's unique position in the solar system and the unusual properties of liquid water that have transformed a hot, barren wilderness into the Blue Planet. The early oceans were the cradle of life on Earth and they still affect life on the planet, controlling weather patterns and major environmental cycles. Life in the oceans encompasses some of the smallest living organisms and the largest animals that have ever existed. Some are familiar from seaside holidays, sailing or diving, but beyond the low tide mark there is an abundance of marine plants and animals that are not so well known and whose biology is not fully understood. Our exploitation of the biological and mineral wealth of the oceans is exacting a toll on the health of the seas. Understanding how the oceans work and devising means to remedy problems are vital to ourselves and future generations. The latest technology is allowing scientists to explore ever-deeper and in greater detail, and may even result in the exploration of seas on other planets. Our journey through the ocean realms will take us from the shores to the deepest places on the surface of Earth.

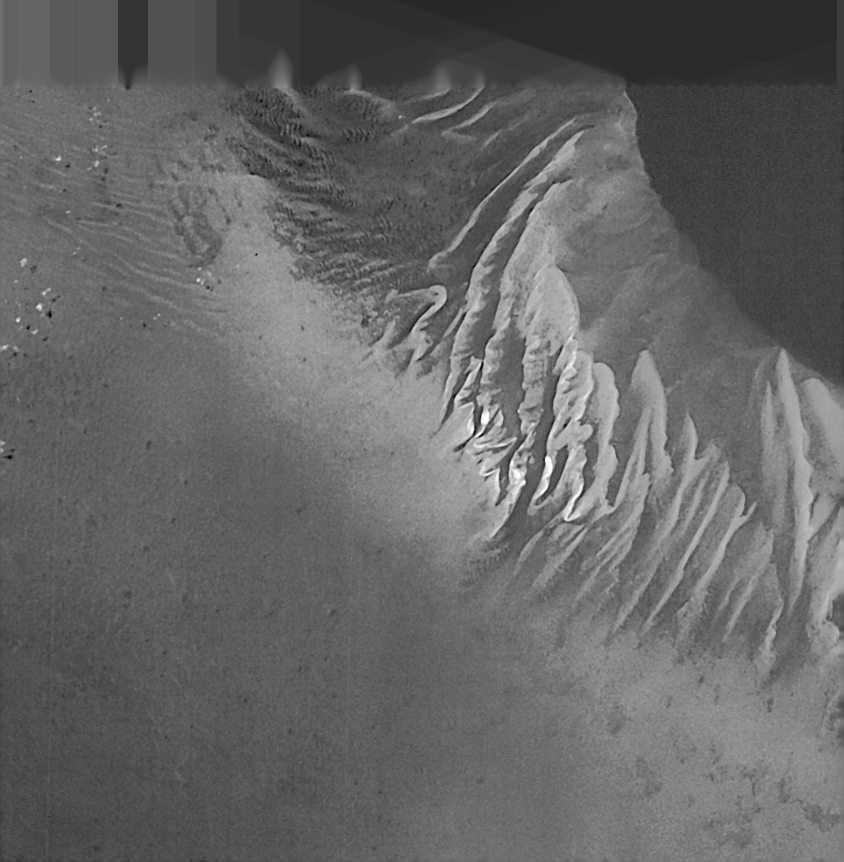

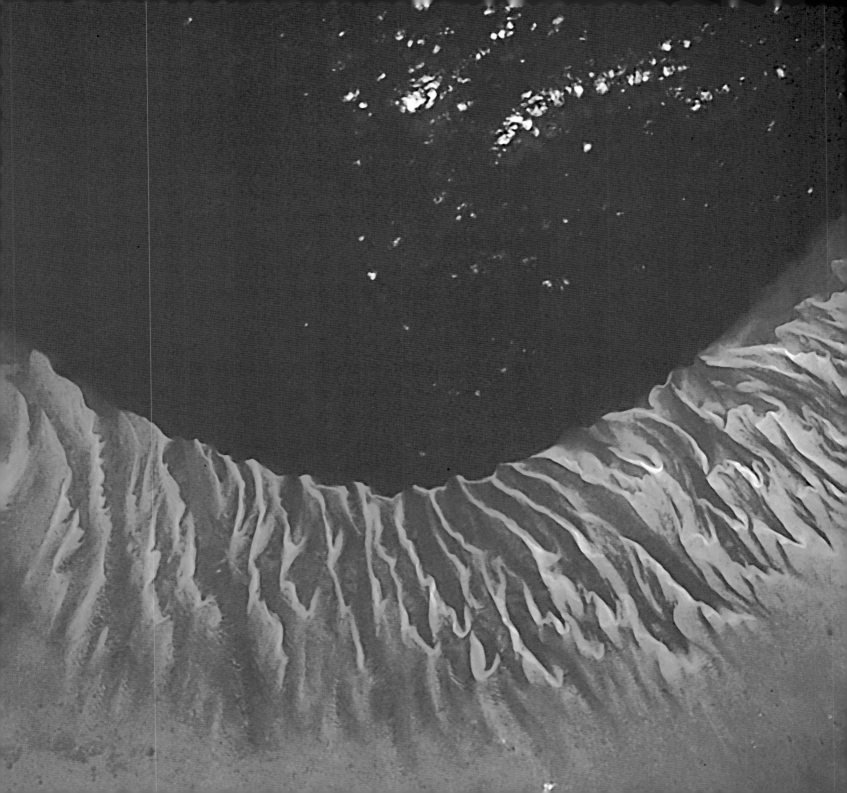

barren wilderness into the only water-covered planet in the Solar System. The oceans still affect life on Earth, controlling weather patterns and impacting upon global environmental cycles.

The blue planet

Planet Earth is unique in the Solar System in that 71 percent of its surface is covered in liquid water. Viewed from space, it is the Blue Planet. The presence of liquid water, and the development of oceans in which life probably formed, are the result of a combination of cosmic factors. The most significant is Earth's position in the Solar System. Earth is the third planet from the Sun and is sufficiently far away from it that water does not instantly turn to vapor. Earth is, however, close enough to be warmed by the Sun so that water and components of our atmosphere are not frozen solid. The size and structure of Earth, as well as its orbit around the Sun, are other crucial factors in producing a water world that harbors life. Earth has sufficient mass to generate a gravitational field that holds on to atmospheric gases as they issue from the surface. The rotation of Earth about its own axis has ensured that our planet is warmed more or less evenly across its entire surface.

THE WATER MOLECULE

A water molecule consists of two hydrogen atoms and a single oxygen atom. The side of the molecule with the hydrogen atoms has a positive electric charge, while the other side is negatively charged. This difference provides an electrical attraction between molecules, known as hydrogen bonding.

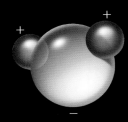

PROPERTIES OF WATER

Although the structure of water is simple, the unique combination of hydrogen and oxygen atoms gives it complex chemical and physical properties that are still not fully understood. For example, water's melting point, 32°F (0°C), and boiling point, 212°F (100°C), are much higher than would be expected by comparison with similar molecules such as hydrogen sulfide and ammonia. Water is unusual in its solid form, ice, because it is less dense than when it is liquid.

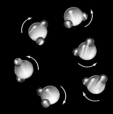
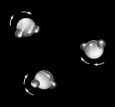
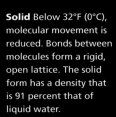

Solid Below 32°F (0°C), molecular movement is reduced. Bonds between molecules form a rigid, open lattice. The solid form has a density that is 91 percent of that of liquid water.

Liquid Sometimes known as the universal solvent, liquid water is important in its ability to dissolve a wide range of substances. It also acts as a medium for biochemical reactions.

Gas To turn liquid water to gas, a great deal of energy is required to make the water molecules move sufficiently fast so they are not held together by electrostatic forces.

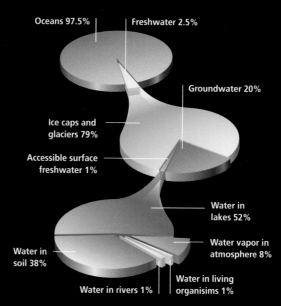

Oceans 97.5%   Freshwater 2.5%

Groundwater 20%

Ice caps and
glaciers 79%

Accessible surface
freshwater 1%

Water in
lakes 52%

Water vapor in
atmosphere 8%

Water in
soil 38%

Water in rivers 1%   Water in living
organisims 1%

DISTRIBUTION OF WATER ON EARTH

The oceans contain 97.5% of the world's water. The remaining 2.5% is distributed between groundwater, ice caps, freshwater lakes and rivers, water in soil and atmospheric water vapor. The largest amount of non-oceanic water is locked up in the ice caps and glaciers (79%) and most of the rest is in groundwater (20%). Only 1% of freshwater is accessible to living organisms, and 1% of this is within living organisms.

↓ **Energy and matter move** around Earth in endlessly repeating cycles. The character and timescales of these cycles have determined the course of evolution of life on Earth and will determine its continuation. The oceans are a key component in these cycles. The hydrological, or water, cycle is a continuous exchange of moisture between the oceans, the atmosphere and the land. Water vapor in the atmosphere falls to the surface as rain or snow. It eventually finds its way back to the sea in rivers, glaciers or subsurface seepages. Water on the sea surface evaporates by solar heating and returns once more to the atmosphere. The cycle then starts again.

Water evaporates
from the sea and
condenses to
form cloud

Rain falls
from cloud

Rivers drain
into the ocean

Inland water
storages are filled

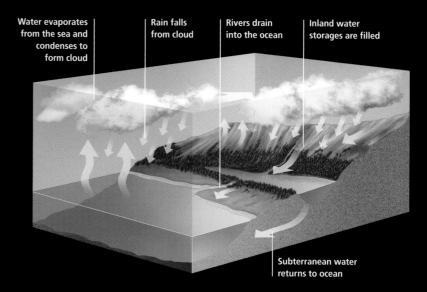

Subterranean water
returns to ocean

Birth of the oceans

The surface of Earth has evolved, during the 4,600 million years of its life, from a totally molten planet to one with a hard skin. Today, the outer skin, or crust, is comprised of either continents or ocean floor. Large, rigid areas of Earth's crust are called tectonic plates and are constantly moving and reshaping the continents and oceans. When Earth's crust solidified sufficiently to enable oceans to form, the water in them came from two sources. The first was steam and other gases vented by volcanic activity deep in Earth's mantle over millions of years. The second source was extraterrestrial. It is estimated that between 5 and 30 icy comets, up to 40 feet (12 m) across, strike the atmosphere every day. There are also satellite observations of much larger so-called space snowballs vaporizing over the Atlantic and adding water to the ocean. Some scientists have suggested that at one stage, the early Earth was entirely covered by water.

↗ **Earth is the only terrestrial planet** whose interior is made up of four components. The crust varies in thickness from about 5 miles (8 km) beneath the oceans to 45 miles (70 km) under the continents. The mantle extends to a depth of 1800 miles (2900 km). The outer core, made of nickel–iron, has a radius of 2200 miles (3500 km). The inner core is 750 miles (1200 km) in radius.

CROSS-SECTION OF EARTH

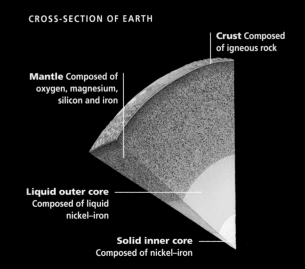

Crust Composed of igneous rock

Mantle Composed of oxygen, magnesium, silicon and iron

Liquid outer core Composed of liquid nickel–iron

Solid inner core Composed of nickel–iron

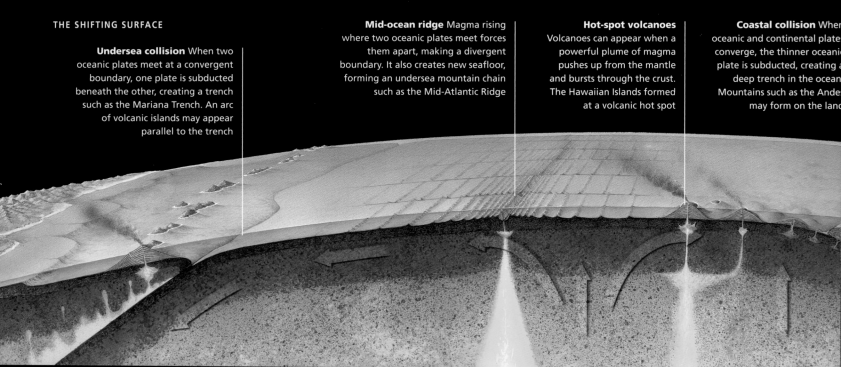

THE SHIFTING SURFACE

Undersea collision When two oceanic plates meet at a convergent boundary, one plate is subducted beneath the other, creating a trench such as the Mariana Trench. An arc of volcanic islands may appear parallel to the trench

Mid-ocean ridge Magma rising where two oceanic plates meet forces them apart, making a divergent boundary. It also creates new seafloor, forming an undersea mountain chain such as the Mid-Atlantic Ridge

Hot-spot volcanoes Volcanoes can appear when a powerful plume of magma pushes up from the mantle and bursts through the crust. The Hawaiian Islands formed at a volcanic hot spot

Coastal collision When oceanic and continental plate converge, the thinner oceani plate is subducted, creating deep trench in the ocean Mountains such as the Ande may form on the lan

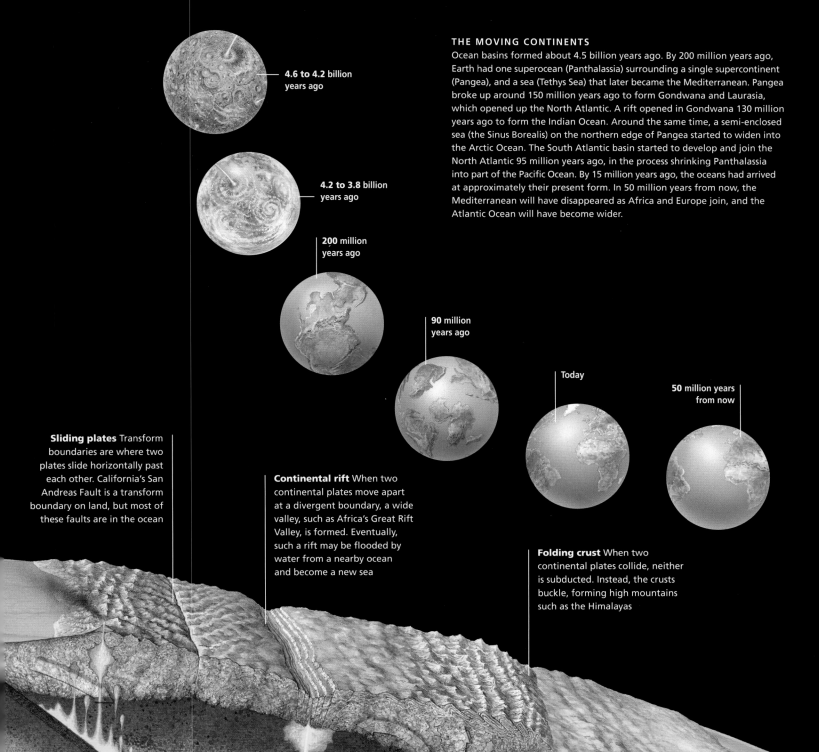

4.6 to 4.2 billion years ago

4.2 to 3.8 billion years ago

200 million years ago

90 million years ago

Today

50 million years from now

THE MOVING CONTINENTS

Ocean basins formed about 4.5 billion years ago. By 200 million years ago, Earth had one superocean (Panthalassia) surrounding a single supercontinent (Pangea), and a sea (Tethys Sea) that later became the Mediterranean. Pangea broke up around 150 million years ago to form Gondwana and Laurasia, which opened up the North Atlantic. A rift opened in Gondwana 130 million years ago to form the Indian Ocean. Around the same time, a semi-enclosed sea (the Sinus Borealis) on the northern edge of Pangea started to widen into the Arctic Ocean. The South Atlantic basin started to develop and join the North Atlantic 95 million years ago, in the process shrinking Panthalassia into part of the Pacific Ocean. By 15 million years ago, the oceans had arrived at approximately their present form. In 50 million years from now, the Mediterranean will have disappeared as Africa and Europe join, and the Atlantic Ocean will have become wider.

Sliding plates Transform boundaries are where two plates slide horizontally past each other. California's San Andreas Fault is a transform boundary on land, but most of these faults are in the ocean

Continental rift When two continental plates move apart at a divergent boundary, a wide valley, such as Africa's Great Rift Valley, is formed. Eventually, such a rift may be flooded by water from a nearby ocean and become a new sea

Folding crust When two continental plates collide, neither is subducted. Instead, the crusts buckle, forming high mountains such as the Himalayas

Spreading seas

Earth's surface is made up of plates moved by forces deep within the planet that are constantly changing shape as they move against one another. In 1912, German meteorologist Alfred Wegener suggested that ocean basins and continents moved slowly over time into their present form. How they moved was not explained until 1960, when American geophysicist Harry Hess suggested that molten magma from Earth boils up to the surface along the crests of the world's mid-ocean ridges. The emerging magma cools, and is pushed down either side of the ridge by the material behind it. The accumulation of this material created the ocean floor, forced out the crust margins, moved continents and widened the ocean basins. Subduction usually occurs when a thin ocean plate meets a thicker continental plate. The ocean plate is forced downward, while the continental plate buckles along the impact zone. Subduction can produce ocean trenches, earthquakes and volcanoes.

SUBDUCTING PLATES

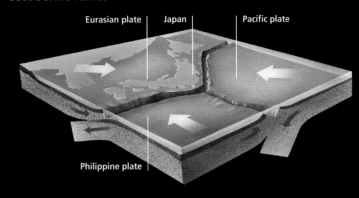

Eurasian plate | Japan | Pacific plate

Philippine plate

↗ **Japan lies at a triple junction** of plates that is the site of intense seismic and volcanic activity. Here, the Pacific plate is subducted beneath the Eurasian and Philippine plates, with the Philippine plate subducted under the Eurasian plate.

Earthquake zone
Prominent hot spot
Convergent margin
Divergent margin
Transform fault
Diffuse or uncertain
Movement direction
Volcanic zone

↑ **A flow of molten lava** pours into the sea on the coast of Hawaii. The lava is cooled by the sea and accumulates on the seafloor.

← **All along mid-ocean** ridges, undersea volcanoes are found at hot spots where plumes of hot magma rise up through the crust. Volcanoes sometimes grow large enough to form volcanic islands such as those on the Hawaiian Ridge.

EVIDENCE FOR SPREADING SEAS

Knowledge of how new ocean crust is formed has been obtained from several different studies of the ocean floors. Traces of Earth's magnetic field left in the magma when it was cooling are used to map the seafloor's magnetic properties. These maps show alternating stripes on either side of the ridges, which indicates that the ocean floors have been laid down over millions of years. In the course of this time, Earth's magnetic field has changed polarity several times. This gives a timescale for seafloor spreading. Sediments over the ocean crust also show that the youngest rocks are close to ridges, while the oldest parts of the ocean floor are situated at the margins. The presence of hot rocks and molten magma that lie under the oceans and well up at the ridges has been detected using sensitive probes placed on the seafloor. These probes indicate that there is much greater heat flow around the mid-ocean ridges than can be found in the rest of the ocean crust, or in continental crustal plates.

Ocean zones

The oceans are an infinitely complex three-dimensional assemblage of structures and processes. They occupy the depressions, or basins, in the surface of Earth that are created by the edges of surrounding continental landmasses. The margins of these basins are an extension of the adjacent landmass, known as the continental shelf, that is covered by relatively shallow seas. Most ocean life and human activity is centered around the continental shelf zone. The outer edge of the continental shelf slopes steeply to the much deeper ocean floor, and the oceanic crustal plate. In the deepest parts of the ocean basins, plate-tectonic processes have shaped the topography of the sediment-covered seafloor. The seafloor ranges from long, rugged ridges to enormous expanses of flat, or abyssal, plains. Even deeper trenches and arcs of volcanic islands are found at geologically active margins. In the water, huge current systems, driven by solar heating, transport heat and chemicals around Earth, affecting the weather over the oceans and on land.

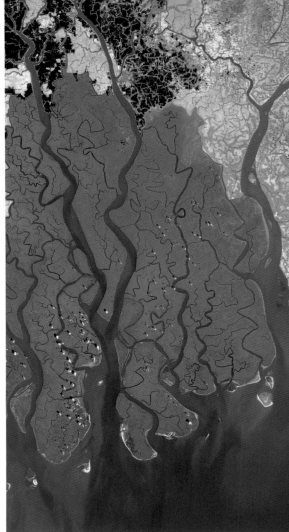

THE ZONES

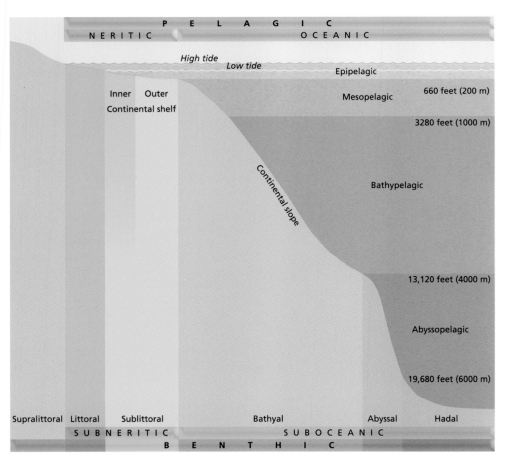

DEFINING THE OCEAN

The most basic distinction in the ocean is made between the features and processes of the water column (pelagic) and those features and processes relating to the seafloor (benthic). Moving horizontally out from the shore to the top of the continental shelf is the shallow area known as the neritic zone. The oceanic realm lies beyond the point of the continental shelf. Within the neritic zone, the shore between the tides is the littoral zone, bordered by the supralittoral above high water and the sublittoral below low tide. The zone from high water to the shelf break is the subneritic. In the water column, the pelagic zone comprises several layers: the surface epipelagic zone; the mesopelagic zone where light from the surface starts to disappear; the dark bathypelagic and abyssopelagic zones; and the hadal zone in the deepest trenches.

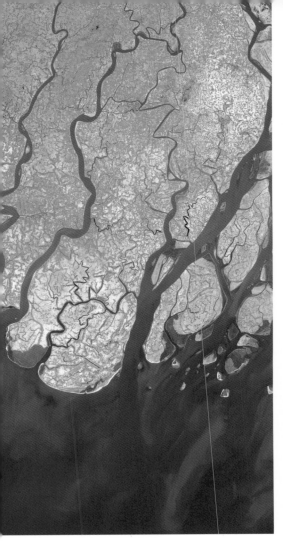

← **Land can intrude** into the sea where major rivers deposit large amounts of sediment in deltas. The River Ganges carries vast quantities of sediment from the land, so that a fan of the material stretches far out over the floor of the northern Indian Ocean.

↓ **Hydrothermal vents** are found along deep ocean ridges. At ridges, the upwelling of magma from deep within Earth creates new seafloor and helps to move the plates that make up the continents and ocean basins.

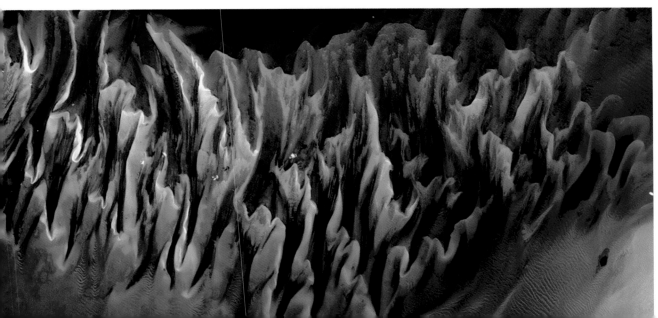

← **These unusual structures** are sand beds in an area of strong tides and currents. The beds were originally laid down where conditions in the water column and the topography of the seafloor slowed the currents carrying the sediments. This change caused the currents to deposit their load. In time, the layers of sediment would have created a firm surface, but the changes in flow have since eroded much of the material.

World ocean

The 71 percent of the world covered in seawater is normally thought of as being divided into four large basins—the Pacific, Atlantic, Indian and Arctic—with smaller subdivisions and adjacent marginal seas. However, a world map shows that all the great oceans are interconnected and exchange water, heat and organisms. Hence, oceanographers often refer to Earth's seawater as the "world ocean." From the south polar view, the Atlantic, Indian and Pacific oceans appear as branches of a single system that extend northward from the Southern Ocean and between the major continental landmasses. The Arctic Ocean, however, is almost landlocked by the Eurasian and North American landmasses, and much of it is covered by permanent ice. The Southern Ocean has been officially recognized since the year 2000 and its northern boundary has been fixed at 60°S. The major oceans are subdivided into smaller seas, gulfs or bays, that are usually defined by obvious geographical boundaries such as the Mediterranean and Black seas. Some areas are delimited by less obvious features such as the Sargasso Sea in the Atlantic Ocean, which is characterized by the mass of seaweed that accumulates. Boundaries between oceans are usually marked by the coastlines of the continental landmasses at their margins, or by major underwater features such as ocean ridges. Not all seas are part of the world ocean—there are large, isolated bodies of saltwater, such as the Caspian and Salton Seas, that are salt lakes.

NORTHERN HEMISPHERE   **SOUTHERN HEMISPHERE**

THE NORTHERN AND SOUTHERN HEMISPHERES
More than two-thirds of Earth's land area is found in the northern hemisphere, while oceans cover at least 80 percent of the southern hemisphere. The distribution of continents and ocean basins shows that many of them are diametrically opposite each other (antipodal), so that continents on one hemisphere are mirrored on the opposing hemisphere by an ocean basin. For example, Antarctica is antipodal to the Arctic Ocean, and Europe is antipodal to the South Pacific Ocean.

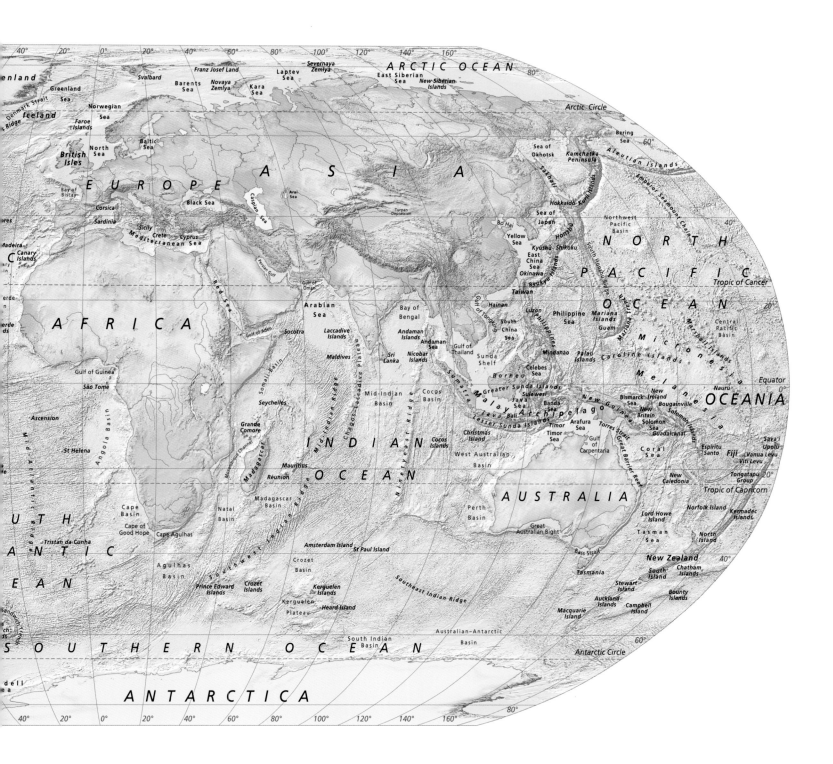

40° 20° 0° 20° 40° 60° 80° 100° 120° 140° 160° 80°

ARCTIC OCEAN

Greenland

Greenland
Sea

Denmark Strait
Iceland
s Ridge

Franz Josef Land
Svalbard
Severnaya
Zemlya

Laptev
Sea

East Siberian
Sea

New Siberian
Islands

Arctic Circle

Norwegian
Sea

Barents
Sea

Novaya
Zemlya

Kara
Sea

Bering
Sea

60°

Faroe
Islands

Baltic
Sea

Sea of
Okhotsk

Kamchatka
Peninsula

Aleutian Islands

British
Isles

North
Sea

Bay of
Biscay

E U R O P E

A S I A

Caspian
Sea

Aral
Sea

Sea of
Japan

Hokkaidō

Kuril Islands

Emperor Seamount Chain

Northwest
Pacific
Basin

40°

Corsica

Black Sea

Turpan
Depression

Bō Hai

N O R T H

Sardinia

Sicily
Crete
Cyprus
Mediterranean Sea

Yellow
Sea

Honshu

Kyūshū Shikoku

East
China
Sea

P A C I F I C

Madeira

C

Canary
Islands

Persian Gulf

Okinawa

Ryukyu Islands

South Honshu Ridge

O C E A N

Tropic of Cancer

Gulf of
Oman

Red Sea

Gulf of
Tonking

Taiwan

Ryukyu Islands

20°

Verde

Gulf of Aden

Arabian
Sea

Hainan

Luzon

Philippine
Sea

Mariana
Islands

Guam

Central
Pacific
Basin

A F R I C A

Socotra

Laccadive
Islands

Bay of
Bengal

Andaman
Islands

Andaman
Sea

Gulf of
Thailand

South
China
Sea

Philippines

Mindanao

Palau
Islands

Caroline Islands

Marshall Islands

s

Gulf of Guinea

Maldives

Sri
Lanka

Nicobar
Islands

Sunda
Shelf

M i c r o n e s i a

São Tomé

Somali Basin

Mid-Indian
Basin

Cocos
Basin

Borneo

Celebes
Sea

Sumatra

M e l a

New
Guinea

Bismarck
Sea

New
Ireland

Bougainville

Nauru

Equator

O C E A N I A

Ascension

Seychelles

Greater Sunda Islands

Sulawesi

Banda
Sea

Malay Archipelago

New
Britain

Solomon Islands

Angola
Basin

Grande
Comore

Java
Sea

Bali

Java

Lesser Sunda Islands

Timor

Arafura
Sea

Torres Strait

Solomon
Sea

Guadalcanal

n e s i a

St Helena

Ninetyeast Ridge

Cocos
Islands

Christmas
Island

Timor
Sea

Gulf of
Carpentaria

Great Barrier Reef

Espiritu
Santo

Sava'i
Upolu

Fiji Vanua Levu
Viti Levu

Mid-Atlantic Ridge

Madagascar

Mauritius

Réunion

I N D I A N

Chagos-Laccadive Plateau

Central Indian Ridge

O C E A N

Cocos
Islands

West Australian
Basin

New
Caledonia

Coral
Sea

Tongatapu
Group

20°

Tropic of Capricorn

Cape
Basin

Natal
Basin

Madagascar
Basin

Perth
Basin

A U S T R A L I A

Lord Howe
Island

Norfolk Island

Kermadec
Islands

S O U T H

Tristan da Cunha

Cape of
Good Hope

Cape Agulhas

Great
Australian Bight

Tasman
Sea

North Island

40°

A T L A N T I C

Agulhas
Basin

Southwest Indian Ridge

Amsterdam Island

St Paul Island

Crozet
Basin

Bass Strait

New Zealand

O C E A N

Prince Edward
Islands

Crozet
Islands

Kerguelen
Islands

Southeast Indian Ridge

Tasmania

South
Island

Chatham
Islands

Stewart
Island

Bounty
Islands

Sandwich Trench

Kerguelen
Plateau

Heard Island

Auckland
Islands

Campbell
Island

ch
ls

Macquarie
Island

Australian–Antarctic
Basin

60°

S O U T H E R N   O C E A N

South Indian
Basin

Antarctic Circle

dell
a

A N T A R C T I C A

40° 20° 0° 20° 40° 60° 80° 100° 120° 140° 160° 80°

Atlantic Ocean

The Atlantic is the world's second-largest ocean, covering approximately one-fifth of Earth's surface. In comparison to the Pacific, the Atlantic is a relatively young ocean, as it formed around 150 million years ago. It is still widening at a rate of approximately 1 inch (2.5 cm) per year because of seafloor spreading at the mid-Atlantic ridge. The mid-Atlantic ridge runs for some 7000 miles (11,000 km) down the center of the Atlantic, from a point north of Iceland, through the equator to Bouvet Island on the edge of the Southern Ocean. Although not as large as the Pacific, the Atlantic has most of the world's great rivers draining into it, since the continental landmasses that surround it generally slope toward it. On its North and South American margins, there are freshwater inputs from the St Lawrence, the Mississippi, the Orinoco, the Amazon and the River Plate. On the west African coasts, the Congo and Niger flow into the central Atlantic. The northeast Atlantic has inputs from the Loire, the Rhine, the Elbe, and the great rivers draining into the Mediterranean, Black and Baltic seas.

The margins of the Atlantic Ocean are far more stable than those of the Pacific. The margins are characterized by broad continental shelves that are the location of many important fishing grounds. The Atlantic shelves also contain substantial oil, gas and other mineral resources.

ATLANTIC OCEAN FACTS	
Area	33.5 million sq miles (86.9 mil. km²)
Average depth	11,828 feet (3605 m)
Maximum depth	28,233 feet (8605 m)
Maximum width	4909 miles (7900 km)
Maximum length	8774 miles (14,120 km)
Coastline length	69,514 miles (111,866 km)
Precipitation (per year)	31 inches (78 cm)
Runoff from land (per year)	8 inches (20 cm)
Evaporation (per year)	41 inches (104 cm)
Ocean water exchange (per year)	2 inches (6 cm)

ATLANTIC COMPOSITION

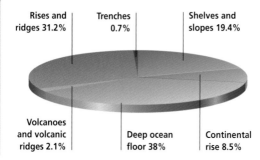

Rises and ridges 31.2%
Trenches 0.7%
Shelves and slopes 19.4%
Volcanoes and volcanic ridges 2.1%
Deep ocean floor 38%
Continental rise 8.5%

LIVING AND MINERAL RESOURCES

The Atlantic shelf seas contain some of the oldest and best-known commercial fisheries. Over the years, these fisheries have produced immense fortunes and dominated the economies of many nations. The cod and herring fisheries in the North Sea and the Newfoundland cod fisheries were the first to develop methods and equipment to allow trawlers to spend long periods at sea, fishing waters far from their ports. Since the 1990s, the collapse of fish stocks in the North Atlantic has been matched by the rise of the South Atlantic as a focus of fishing activity, in particular, the harvesting of squid by Japanese and Korean fleets.

The mineral wealth of the Atlantic has principally been its reserves of oil and gas in its shelf seas. After more that 30 years of production, however, reserves are nearing complete exhaustion. Oil exploration in the North Atlantic is now looking to deep waters west of the Shetland Islands. In the South Atlantic, the Falkland Plateau is also thought to contain significant reserves.

Indian Ocean

The Indian Ocean is the third largest of the world's oceans and occupies just under 20 percent of the total world ocean area. Its deepest point, at 24,460 feet (7455 m), is in the Java Trench. The Indian Ocean was formed around 140 million years ago when the continent of Gondwana broke apart, separating the Indian, African and Antarctic landmasses. The Red Sea, which is a narrow offshoot of the Indian Ocean, is still widening because it lies over a spreading ridge that is slowly separating Africa from the Arabian Peninsula. The Indian Ocean contains the largest deposit of river sediments on Earth because two of the largest rivers in the world, the Indus and Ganges, flow into it on either side of the Indian subcontinent. The Ganges fan stretches 1240 miles (2000 km) southward from its delta over the floor of the Bay of Bengal.

The current systems in the Indian Ocean are unique in that they change direction twice a year. Currents in all other oceans flow in the same direction year round. In winter, monsoon

winds force currents to flow toward Africa, and in summer, the prevailing winds move currents toward India. The deep water of the Indian Ocean has its origins in the Persian Gulf and Red Sea. As a consequence, most of the water column has a high salinity and low oxygen content. The most famous inhabitant of the Indian Ocean is the coelacanth fish, which was thought to be extinct until it was recognized in fishermen's catches in 1938.

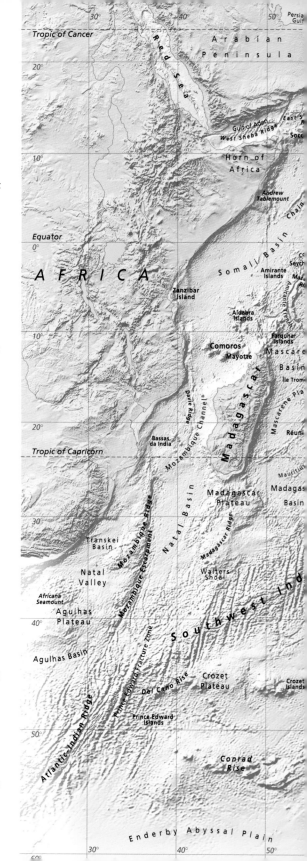

INDIAN OCEAN FACTS	
Area	26.9 million sq miles (70 mil. km²)
Average depth	12,645 feet (3854 m)
Maximum depth	24,460 feet (7455 m)
Maximum width	6338 miles (10,200 km)
Maximum length	5841 miles (9400 km)
Coastline length	41,339 miles (66,526 km)
Precipitation (per year)	40 inches (101 cm)
Runoff from adjoining land (per year)	3 inches (7.5 cm)
Evaporation (per year)	54 inches (138 cm)
Ocean water exchange (per year)	12 inches (30 cm)

INDIAN COMPOSITION

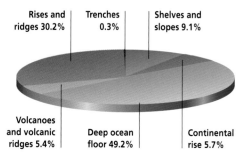

Rises and ridges 30.2%
Trenches 0.3%
Shelves and slopes 9.1%
Volcanoes and volcanic ridges 5.4%
Deep ocean floor 49.2%
Continental rise 5.7%

LIVING AND MINERAL RESOURCES

Although the Indian Ocean is almost the same size as the Atlantic, it yields only about one-fifth of the quantity of fishes harvested annually from either the Atlantic or Pacific. This difference is thought to be a result of a combination of factors. Most fishing is carried out in small, nearshore fisheries that supply the needs of local coastal communities rather than being a large-scale activity for the global market. Fisheries scientists believe that the lack of extensive continental shelf seas, and the seasonal reversal in wind patterns, limit the surface production of food species that, in turn, limits the size of fish populations.

Exploitation of the mineral wealth of the Indian Ocean has concentrated on the oil and gas fields of the Persian Gulf and, more recently, off western Australia. However, there are also extensive deposits of other minerals, such as manganese nodules and offshore phosphate deposits, that can be used as fertilizers. Deposits of the raw form of potash fertilizers have long been extracted along the east coast of South Africa.

ASIA

Tropic of Cancer

Arabian Sea

Arabian Basin

Laccadive Islands

Cape Comorin

Sri Lanka

Maldives

Chagos Archipelago

Diego Garcia

Murray Ridge

Owen Fracture Zone

Chagos-Laccadive Ridge

Chagos Trench

Mid-Indian Ridge

Ceylon Plain

Mid-Indian Basin

Ganges Cone

Bay of Bengal

Andaman Islands

Andaman Sea

Andaman Basin

Nicobar Islands

Ninetyeast Ridge

Cocos Basin

North Keeling Island

Cocos Islands

Home Island

Sumatra

Strait of Malacca

Investigator Ridge

Java Ridge

Christmas Island

Java Trench 23,376ft (7125m)

Gulf of Tongking

Hainan

South China Sea

Gulf of Thailand

Sunda Shelf

Greater Sunda Islands

Borneo

Makassar Strait

Java Sea

Java

Philippines

Celebes Sea

Sulawesi

Seram

Bali

Flores Sea

Lesser Sunda Islands

Sumbawa

Lombok Basin

Timor

Banda Sea

Equator

New Guinea

Arafura Sea

Timor Trough

Timor Sea

Melville Island

Arafura Shelf

Sahul Shelf

Torres Strait

Cape York

10°

Coral Sea

Gulf of Carpentaria

Cape Leveque

North Australian Basin

Osborn Plateau

Gascoyne Plain

Exmouth Plateau

Rowley Shoals

Wharton Basin

Wallaby Plateau

Cuvier Plateau

Cuvier Basin

Tropic of Capricorn

INDIAN

OCEAN

Argados Carajos Islands

Rodrigues Island

East Indiaman Ridge

Batavia Seamount

Golden Dragon Seamount

Broken Plateau

Houtman Ridge

Perth Basin

Naturaliste Plateau

AUSTRALIA

Amsterdam Fracture Zone

Naturaliste Fracture Zone

Cape Leeuwin

Diamantina Fracture Zone

Diamantina Deep 21,660ft (6602m)

Amsterdam Island

St Paul Island

Great Australian Bight

South Australian Basin

King Island

Bass Strait

Flinders Island

Tasman Sea

40°

rozet Basin

Kerguelen Islands

Southeast Indian Ridge

South Australian Plain

Tasmania

South East Cape

South Tasman Rise

50°

MacDonald Islands

Heard Island

Kerguelen Plateau

Indian-Antarctic Ridge

SCALE 1:47,000,000
Miller Projection

Banzare Seamount

South Indian Basin

Pacific Ocean

The Pacific is the world's largest ocean, covering about one-third of Earth's surface. It was once even bigger, when it was part of the early great ocean of Panthalassia that surrounded the ancient supercontinent of Pangea. Plate tectonics have opened up the other ocean basins at the expense of the Pacific, but it is still the deepest ocean. The Pacific has a greatest average depth of 13,127 feet (4001 m) and contains the deepest point in the ocean at 36,201 feet (11,034 m) in the Mariana Trench. At its widest point, the Pacific is 11,185 miles (18,000 km) across and reaches almost halfway around Earth. Earth's highest mountain, measured from base to peak, is Mauna Kea on Hawaii. This volcano rises from the Pacific's floor to 38,184 feet (9200 m). The plate margins of the Pacific are extremely active. The subduction of huge slabs of oceanic crust into the underlying mantle produces violent earthquakes and intense volcanic activity in a zone that runs around the edge of the Pacific from New Zealand to southern Chile. This zone is known as the Pacific Ring of Fire.

The immense distances between the more remote islands of the Pacific has meant that they have been isolated from human activities until relatively recent times. This isolation has brought about many unique biological communities. It was the study of the isolated Galapagos flora and fauna that led Charles Darwin to formulate his ideas on evolution.

PACIFIC OCEAN FACTS	
Area	65.6 million sq miles (169.8 mil. km²)
Average depth	13,127 feet (4001 m)
Maximum depth	36,201 feet (7455 m)
Maximum width	11,185 miles (18,000 km)
Maximum length	8637 miles (13,900 km)
Coastline length	84,301 miles (135,663 km)
Precipitation (per year)	48 inches (121 cm)
Runoff from adjoining land (per year)	2 inches (6 cm)
Evaporation (per year)	45 inches (114 cm)
Ocean water exchange (per year)	5 inches (13 cm)

PACIFIC COMPOSITION

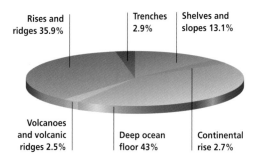

Rises and ridges 35.9%

Trenches 2.9%

Shelves and slopes 13.1%

Volcanoes and volcanic ridges 2.5%

Deep ocean floor 43%

Continental rise 2.7%

LIVING AND MINERAL RESOURCES

The Pacific's two main commercial fishing sites are in the shallow seas of the North Pacific and in the upwelling zones along the coast of South America. In the colder northern waters, there are fisheries for cod, sea bass and various flatfish species. In the upwelling zone, the fisheries are almost entirely for anchovy, huge quantities of which are converted to fish meal each year for livestock feed. Shellfishes are also harvested throughout the Pacific with large-scale fisheries for crabs, lobsters and shrimp in the

Yellow and South China seas and around Australia. The famed Alaskan king crab is caught off Alaska and the Aleutian Islands.

In terms of mineral wealth, the Pacific has enormous fields of metal-rich manganese nodules that have yet to be collected on a commercial scale. However, other minerals, such as iron deposits off Japan, are extracted. The recent discovery of oil and gas in the Malay archipelago has stimulated oil and gas prospecting around the rim of the Pacific.

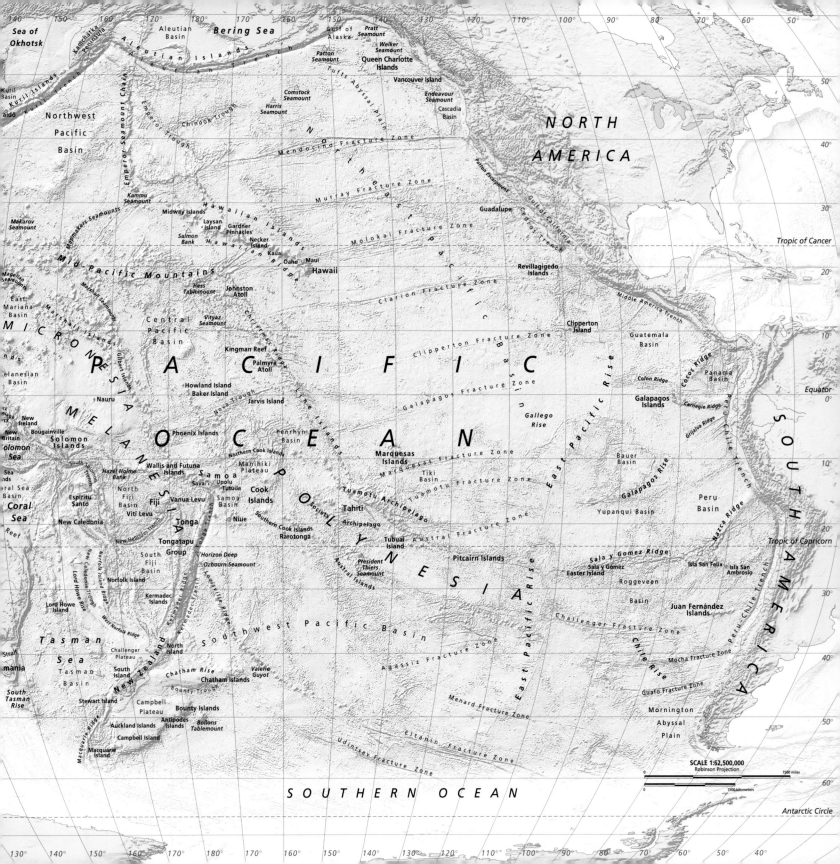

Southern Ocean

The Southern Ocean did not officially exist until 2000, when the International Hydrographic Organization formally defined its boundaries for the purposes of navigation. It is now recognized as the ocean between 60°S and the continent of Antarctica. The choice of 60°S as a boundary for the Southern Ocean was not completely arbitrary, since it closely approximates to the southern edge of the Antarctic Circumpolar Current. This current flows clockwise around Antarctica and effectively isolates the surface waters of the Southern Ocean from oceans to the north. The continental shelf in the Southern Ocean is generally deep and narrow, covered by 1200 to 1600 feet (365 to 490 m) of water, compared to an average depth of 600 feet (180 m) in the Atlantic shelf. Its deepest point is 23,736 feet (7235 m) in the South Sandwich Trench. It is noted for ferocious weather and high waves. At some latitudes, it is possible in theory for the waves to have an infinite fetch (the distance traveled with no obstruction), since they flow around the world without touching land.

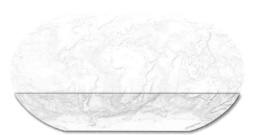

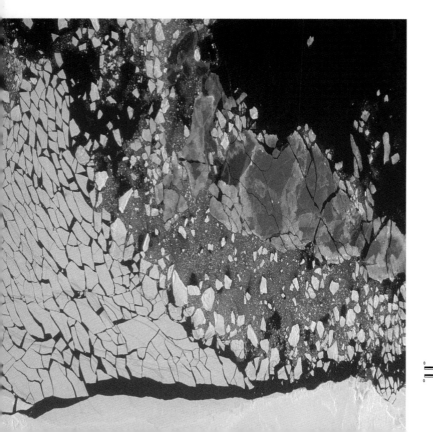

SCALE 1:57,662,700
Lamberts Azimuthal Equal Area Projection

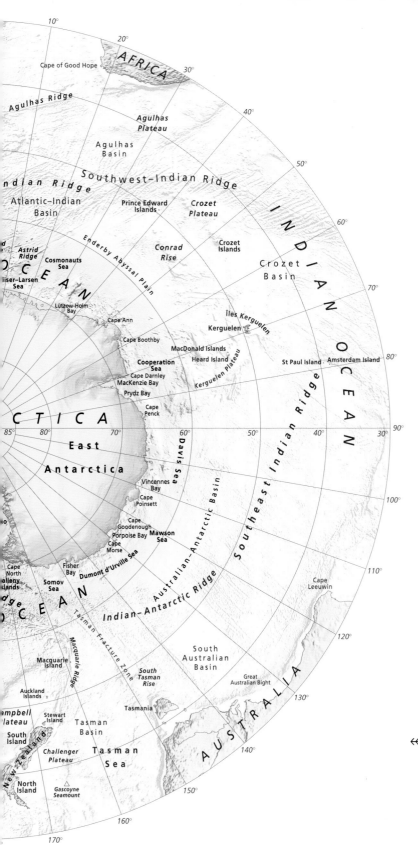

AFRICA

Cape of Good Hope

Agulhas Ridge

Agulhas Plateau

Agulhas Basin

Southwest-Indian Ridge

...dian Ridge

Atlantic-Indian Basin

Prince Edward Islands

Crozet Plateau

INDIAN OCEAN

Astrid Ridge

Cosmonauts Sea

Conrad Rise

Crozet Islands

...liser-Larsen Sea

Crozet Basin

Enderby Abyssal Plain

Lützow-Holm Bay

Cape Ann

Îles Kerguelen

Kerguelen

Cape Boothby

MacDonald Islands

Cooperation Sea

Heard Island

St Paul Island

Amsterdam Island

Cape Darnley

MacKenzie Bay

Kerguelen Plateau

Prydz Bay

Cape Penck

...CTICA

East Antarctica

Davis Sea

Southeast Indian Ridge

Vincennes Bay

Cape Poinsett

Cape Goodenough

Porpoise Bay

Mawson Sea

Cape Morse

Fisher Bay

Dumont d'Urville Sea

Indian–Antarctic Basin

Cape North ...lany Islands

Somov Sea

Indian–Antarctic Ridge

...OCEAN

Cape Leeuwin

Tasman Fracture Zone

Macquarie Ridge

Macquarie Island

South Australian Basin

South Tasman Rise

Great Australian Bight

Auckland Islands

AUSTRALIA

...mpbell ...lateau

Stewart Island

Tasman Basin

Tasmania

South Island

Challenger Plateau

Tasman Sea

New Zealand

North Island

Gascoyne Seamount

SOUTHERN OCEAN FACTS	
Area	7.85 million sq miles (20.3 mil. km²)
Average depth	14,450 feet (4500 m)
Maximum depth	23,736 feet (7235 m)
Maximum width	1678 miles (2700 km)
Maximum length	13,360 miles (21,500 km)
Coastline length	11,165 miles (17,968 km)

↑ **Ships carrying supplies,** scientists and tourists to Antarctica tend to converge on the west coast of the Ross Sea (*pictured*), which is the most predictably open area during the Antarctic summer.

↤ **Pack ice in McMurdo Sound breaks up** as winter turns to spring. Access to Antarctica is blocked for much of the year by the pack ice that forms around its coastline at the start of March. The ice gradually thickens and extends out from the coast until spring. In January, the slight rise in temperature causes the pack ice to break away in large pieces that are further broken down by wave action.

Arctic Ocean

The Arctic Ocean is the smallest ocean and the shallowest one, with an average depth of only 4690 feet (1430 m). It is almost completely surrounded by the continental landmasses of Eurasia and North America. The continental shelf in the Arctic Ocean is exceptionally wide, taking up approximately 50 percent of the total seafloor. Since the discovery of oil in Alaska in the 1960s, the Arctic's continental shelf has been the focus of major mineral prospecting. The North Pole lies within the Arctic Ocean. Unlike the South Pole, the North Pole it is not sited on a landmass, but is in a region of sea ice that can be 164 feet (50 m) thick in winter and as little as 6½ feet (2 m) in summer. In 1958, the submarine USS *Nautilus* was able to surface through the ice at the North Pole and prove that there was no Arctic continent hidden beneath the ice. The permanent ice of the Arctic Ocean is referred to as polar ice.

The ice that forms in winter around its edges is called pack ice. "Fast ice" forms between continental shores at the margin and then joins with the pack ice.

SCALE 1:32,650,000
Lamberts Azimuthal Equal Area Projection

0   600 miles

0   600 kilometers

ARCTIC OCEAN FACTS	
Area	5.44 million sq miles (14 mil. km²)
Average depth	4690 feet (1430 m)
Maximum depth	18,456 feet (5625 m)
Maximum width	1988 miles (3200 km)
Maximum length	3107 miles (5000 km)
Coastline length	28,205 miles (45,389 km)

↑ **The sinuous curves** in this enormous ice sheet mark the tracks of glaciers near the sea around Baffin Bay, Canada. The white dots in the sea are icebergs that have separated from the glaciers.

← **Icebreakers have strengthened** hulls that allow them to force their way through leads—thin stretches of open water that open up in the Arctic pack ice. Since the early 1990s, a number of icebreakers have been converted to carry passengers. Today, tourists travel in great comfort to see the spectacular glacial scenery and wildlife, causing little disturbance to this fragile habitat.

Seawater and salinity

The saltiness of seawater is referred to as its salinity. Measurements of salinity are of great importance because salinity affects innumerable physical, chemical and biological processes. The presence of large amounts of dissolved salts in seawater even affects its density; seawater is usually 1.03 times heavier than the same volume of freshwater at the same temperature. The average salinity of the oceans is 35 (salinity is unitless), varying between 33 and 37. This variance depends on the balance between evaporation, which increases salinity, and the effect of diluting rainfall. In partially enclosed seas, salinity ranges are more extreme. For example, in the Red Sea, evaporation produces salinities around 40, but in the Baltic Sea, rivers dilute surface salinity to 7.

→ **The salinity of the Dead Sea** is approximately 10 times that of seawater. The water is so dense, it is impossible for a human body to sink in it.

↓ **Local evaporation** by the Sun may increase surface salinity. However, wind and wave action mix the surface with deeper layers, restoring salinity to normal.

↓ **The composition of seawater.** Percentages refer to the weight of each component. Dissolved salts account for only 3.53 percent of the weight of seawater. Chloride and sodium are the major components of the dissolved salts; they represent more than 85 percent of the total weight of salts.

Chloride 55.04%
Sodium 30.61%
Sulfate 7.68%
Magnesium 3.69%
Potassium 1.10%
Others 0.72%

Dissolved salts 3.53%

Pure H_2O 96.47%

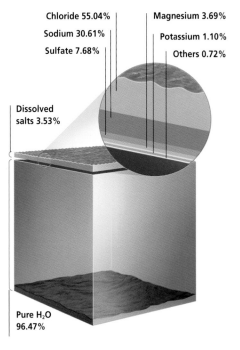

ORIGINS OF THE OCEAN'S SALINITY

Approximately 4.3 billion years ago, Earth entered a 10-million-year period of intense rainstorms. These not only cooled the surface, but also washed minerals from the hot rocks and picked up gases from the atmosphere (see illustration, *right*). As a result, the ocean that formed was a complex solution of salts, rather than freshwater. In present-day oceans, there is a balance between salt inputs and losses, so the salinity remains almost constant.

↑ **The sediment-laden** freshwater (*top right*) is held against the coast (*bottom right*) and does not mix with much denser seawater (*left*), sometimes referred to as a salt plug. The lack of mixing produces a visible boundary between the two water masses.

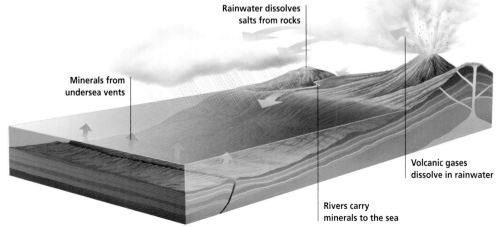

Rainwater dissolves salts from rocks

Minerals from undersea vents

Volcanic gases dissolve in rainwater

Rivers carry minerals to the sea

Solar and lunar tides

The rhythmic daily rise and fall of the sea has been known since earliest times, but it has only been in the last 350 years that we have come to understand the generation and complexity of the tides. Tides are generated by the interaction of the gravitational pulls of the Sun and Moon on Earth, causing the oceans to wash to and fro, as if they were in a giant bowl. The basic celestial rhythm of the tides is altered and made more complex by differences in water depth, the shape of the adjacent landmasses and the rotation of Earth. For example, the shape of the Bay of Fundy, on the east coast of Canada, amplifies the tides so that the area has the greatest tidal range in the world at 56 feet (17 m). The dominant tidal cycles are related to lunar cycles and so take place over lunar days of 24 hours 50 minutes, and change from greatest tidal range (spring tides) to minimum tidal range (neap tides) twice every lunar month of 28 days. There are also longer-term cycles hidden in the tidal charts that have been revealed by the mathematical analyses used to predict tides. Tides were predicted by intricate mechanical calculators but since the 1960s computers have taken over the complex calculations.

SPRING TIDE

Sun

Sun

↓ **The sand flats on Whitsunday Island** are exposed at low water. This part of the Great Barrier Reef has some of the greatest tidal ranges on the east coast of Australia, some reaching a maximum of 29 feet (9 m). Narrow passages on the reef retard the normal semidiurnal tide that is then concentrated into the channels that carry it to the coast, producing a large tidal flow.

TIDAL PATTERNS ACROSS THE WORLD

The simplest tidal pattern is the semidiurnal (twice daily) tide, with high water alternating with low water every 6 hours 12 minutes. In some places, there is just a single tide per day (diurnal) with high and low water separated by 12 hours 25 minutes. Finally, some coasts experience a mixed tidal regime where there are two tides per day, but with a large difference in the height of the tide between consecutive high-water phases. The Atlantic and Indian oceans generally have semidiurnal tides with two equal tides per day. In the Pacific Ocean, there are two tides per day but one is normally much larger than the other.

■ Semidiurnal tides ■ Diurnal tides ■ Mixed tides

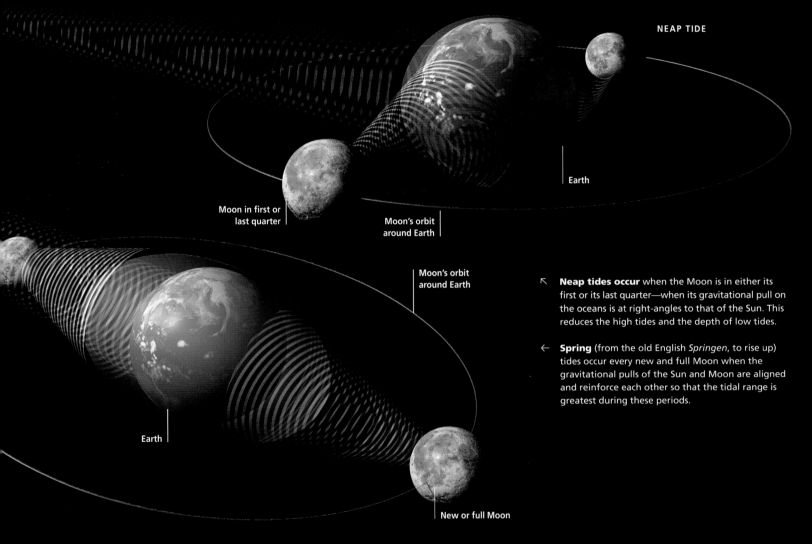

NEAP TIDE

Earth

Moon in first or
last quarter

Moon's orbit
around Earth

Moon's orbit
around Earth

Earth

New or full Moon

↖ **Neap tides occur** when the Moon is in either its
first or its last quarter—when its gravitational pull on
the oceans is at right-angles to that of the Sun. This
reduces the high tides and the depth of low tides.

← **Spring** (from the old English *Springen*, to rise up)
tides occur every new and full Moon when the
gravitational pulls of the Sun and Moon are aligned
and reinforce each other so that the tidal range is
greatest during these periods.

TIDE CURVES FOR THE THREE COMMON TYPES OF TIDES

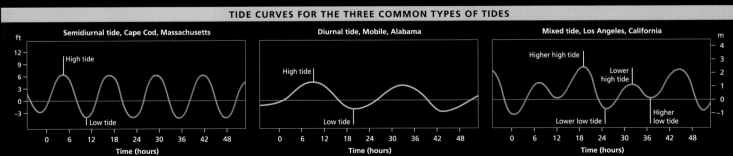

Semidiurnal tide, Cape Cod, Massachusetts

ft
12
9
6
3
0
-3

High tide

Low tide

0   6   12   18   24   30   36   42   48
Time (hours)

Diurnal tide, Mobile, Alabama

High tide

Low tide

0   6   12   18   24   30   36   42   48
Time (hours)

Mixed tide, Los Angeles, California

Higher high tide

Lower
high tide

Lower low tide

Higher
low tide

m
4
3
2
1
0
-1

0   6   12   18   24   30   36   42   48
Time (hours)

Surface currents

Surface currents are divided into open-ocean currents and boundary currents. The three main oceanic basins have similar current patterns. In equatorial regions, the currents follow surface winds and normally flow westward in north and south equatorial currents, driven by the trade winds. These currents are separated by an eastward-flowing equatorial current that coincides with the doldrums. In the high latitudes of the northern hemisphere, there are eastward-flowing surface currents in the North Pacific and North Atlantic, matched in the southern hemisphere by the Antarctic circumpolar current. In each hemisphere interactions between eastward- and westward-flowing currents cause large gyres in subtropical latitudes. Boundary currents flow along the edge of continental margins.

SURFACE CURRENT SYSTEMS

→ Cool current
→ Warm current

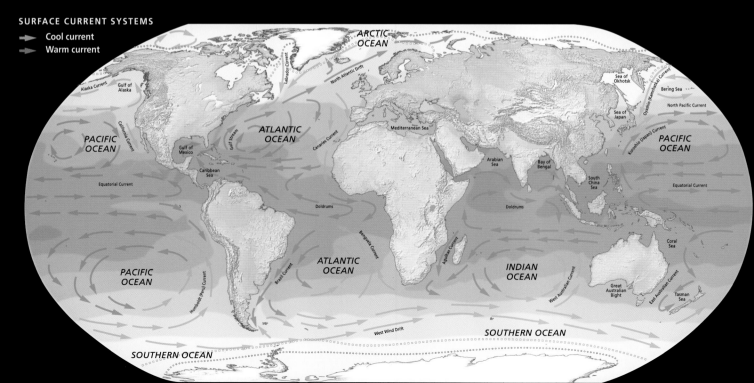

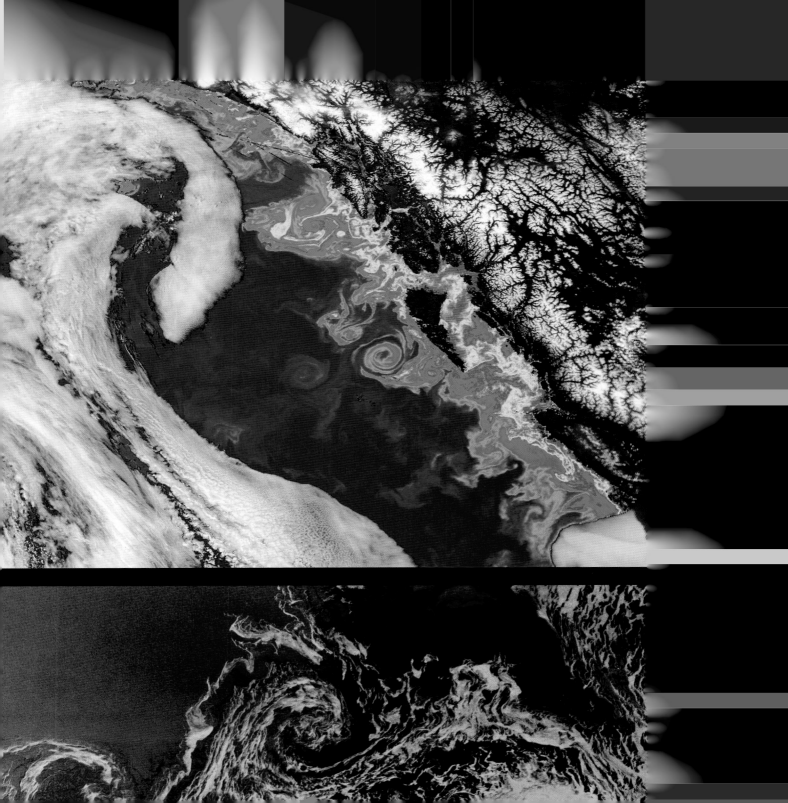

Deep water currents

Surface currents affect only the uppermost 10 percent of the oceans. Most water movements are much bigger, deeper and slower. Deep ocean currents, usually referred to as thermohaline circulation, are not affected by surface winds; they are driven by changes in the density of water that are produced by variations in water temperature and salinity. Thermohaline circulation describes the movement of huge water masses that do not mix easily when they come into contact and can take centuries to complete their circulation around the global ocean. The water masses can be divided into five main types: surface waters down to 660 feet (200 m); central water extending from the bottom end of the surface water to where there is no further drop in temperature (bottom of the thermocline); intermediate water that extends down to 5000 feet (1500 m); deep water that is below intermediate water and is not in contact with the bottom, extending down to 13,000 feet (4000 m); bottom water in contact with the seafloor. The systematic mapping of the oceans' currents over the last 150 years has shown how they are generated and their importance in transferring heat around the world, thereby controlling our planet's climate.

Tidal currents that create the whirlpools off the coast of Saltstraumen, in northern Norway, are so powerful that they produce vertical transport that extends into deep bottom water. The vertical motion is strong enough to bring up deep water fishes so quickly that they are killed as their gas bladders burst.

Antarctic Bottom Water (ABW) is formed mainly in the Weddell Sea. Its salinity and low temperature make it the densest water in the world's oceans. Between 706 and 1766 million cubic feet (20 and 50 million m³) of ABW forms every second.

THE GREAT OCEAN CONVEYOR

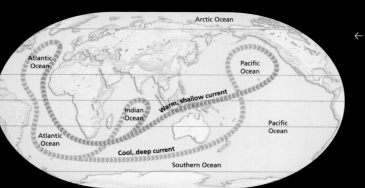

← **The Great Ocean Conveyor** is the name sometimes applied to the deep circulation that links the Pacific, Atlantic and Indian oceans. It is slow-moving with water taking up to 1000 years to circulate, but a billion cubic feet (30 million m³) of water enter the Conveyor every second. This deep thermohaline current system is vital to life in the deep ocean since it carries oxygen down from the surface layers.

WATER MOVEMENT
Seawater becomes denser when it is cold and its salinity is increased. Multiplying this up to ocean scales generates enormous forces that can move immense volumes of water. Cooling at high latitudes causes denser cold water to sink and travel slowly away from the poles as deep and bottom water masses, eventually rising and returning in the upper layers in counter currents.

THERMOHALINE CIRCULATION

Heating

Cooling

Surface flow

Thermocline

Equatorial regions

Deep spreading

Polar regions

Sinking

The Gulf Stream and North Atlantic drift

The Gulf Stream is the biggest and best-known western boundary current and originates in the Gulf of Mexico, flowing northward along the eastern coast of North America at 4 miles per hour (6.5 km/h) and extending down 1500 feet (450 m) from the surface. The flow of water is estimated to be 1 billion cubic feet (30 million m³) per second as it passes through the Straits of Florida—this is about 300 times the normal flow of the Amazon. The Gulf Stream starts to move eastward off Cape Hatteras and meets the cold Labrador Current around the Grand Banks, giving rise to the notorious fogs of this area. As it moves into the Atlantic, the Gulf Stream becomes less defined and is called the North Atlantic Current or drift. The North Atlantic drift divides in the middle of the ocean, one branch moving south, the other carrying warm water to the shores of northwest Europe.

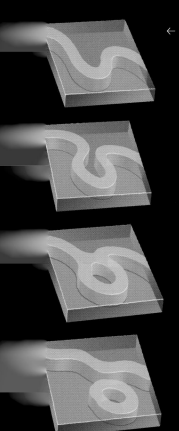

← **The Gulf Stream develops** meanders as it flows that may "bud off" as either cold, or warm-water core eddies, depending on the temperature of the water trapped within them. If the meander cuts off to the north of the main stream, the eddy will trap warm Sargasso Sea water and rotate clockwise. If it is cut off to the south, cold slope water is trapped and the eddy circulates anticlockwise. Eddies can be up 200 miles (320 km) across and last for up to three years, before merging again with the main current.

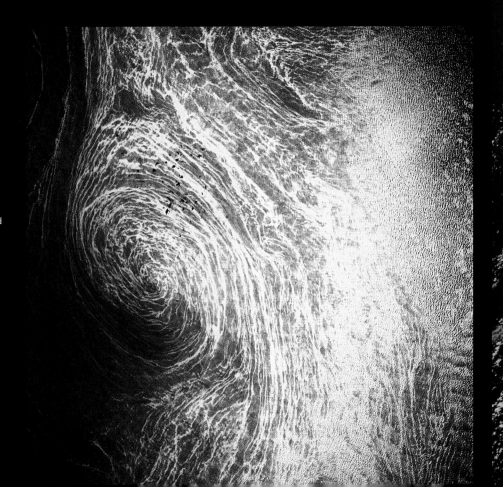

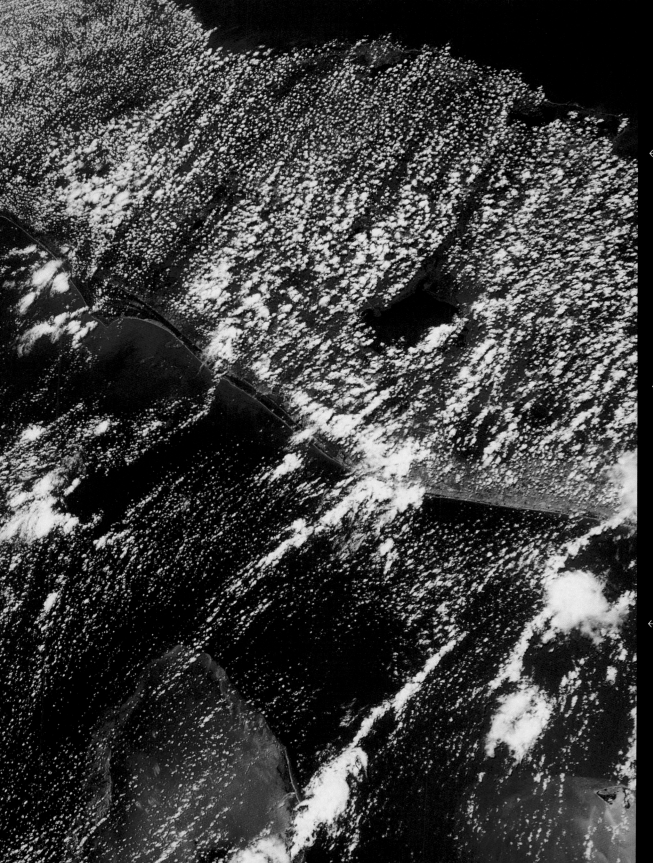

← **Turtles originating from** the Caribbean and the Gulf of Mexico are known to migrate out into the Atlantic, carried by the Gulf Stream. The vast majority are carried south where the North Atlantic drift divides, but some are carried north and end up stranded on the Atlantic coasts of Europe.

← **The boundary between** the Gulf Stream and the surrounding waters is most marked as it flows north along the Florida coast. The water within the current is usually warm, clear and blue because of its lack of nutrients. The water around it is often cloudy and green because of phytoplankton.

← **The final stages in the formation** of a Gulf Stream eddy, just before separation from the main current, were photographed from the space shuttle Endeavour, as it passed over the North East Atlantic. The bright sunshine highlighted the ridges created in the surface of the water by the intense flows.

El Niño | La Niña

This oceanographic and weather phenomenon
is centered around the western coasts of Central
and South America. It takes its name from the
Spanish for the Christ child, since it commences
around Christmas every three to eight years.
In non-El Niño years, the trade winds drag huge
volumes of surface water westward, away from
the coast. This allows the cold, nutrient-rich
waters of the Peru Current to reach the surface
near the coast. These waters support rich supplies
of plankton that are vital to sustaining the local
fisheries and vast numbers of seabirds. In an
El Niño event, for reasons that are still not clear,
the trade winds slacken and warm water remains
at the coast. This water blocks the upwelling of
the Peru Current so that there is no significant
plankton production. Fishes and birds either die
of starvation or go elsewhere. In recent years,
it has been realized that these events also cause
changes in global weather patterns. In the severe
El Niño event of 1997–98, there were exceptional
tornadoes in the American southwest because
of the extra warming of air masses by the coastal
warm water. Droughts in Papua New Guinea,
Hawaii and southwest Africa occurred because
the normal flow of moist, rain-producing air in
the trade winds was blocked.

LA NIÑA

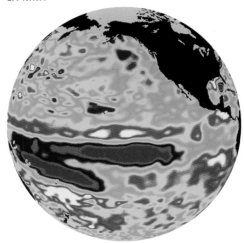

↑ **Floods from the El Niño storm** of 1983
encircle a car parked on a shopping street in
Laguna Beach, California, USA. El Niño events
produce changes in global weather patterns,
causing massive floods in some areas and
catastrophic droughts in others.

← **La Niña is the return** to upwelling conditions
along the coast of Peru and Chile. Following
the 1997–98 El Niño event, a large body of cold
water (purple) was brought to the surface by
the return of the trade winds.

→ **The collapse of fish stocks,** such as anchovy,
salmon and tuna, that followed the El Niño event
in 2003 affected marine wildlife well beyond the
upwelling zone. In a rare event, starving sea lions
came inshore in search of food at Monterey Bay
and along other parts of the Californian coast.

March 17

March 23

April 6

← **The development of the 1997** El Niño event was followed closely by weather satellites. These three Earth graphics show the main stages of the event. It began by suppressing the upwelling of cold water off Peru, which brought warm, central Pacific surface water to the coast of South America. The difference between average sea heights and those caused by the warm water of the El Niño event are shown from purple (below normal level) through blue, green, yellow, red to white (above normal level). El Niño appears as the white patch that moves eastward along the equator across the Pacific Ocean.

OCEAN TEMPERATURES

Average ocean temperature during a La Niña event shows cold surface coastal water off South America. During an El Niño event, warm water is no longer drawn away from the coast, raising the ocean temperature. Coastal water was colder than normal during La Niña and warmer during El Niño.

AVERAGE OCEAN TEMPERATURES

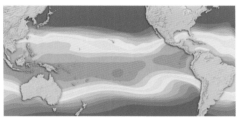

La Niña January–March 1989

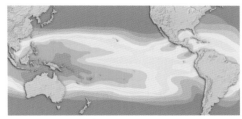

El Niño January–March 1998

64°F　　75°F　　86°F
18°F　24°F　30°F

OCEAN TEMPERATURE ANOMALIES

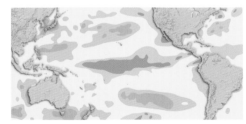

La Niña January–March 1989

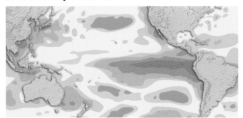

El Niño January–March 1998

−5.4°F　0°F　+5.4°F
−3°F　0°F　+3°F

Wave generation

Waves are generated by wind blowing over the surface of the sea. Despite appearances, a wave is not a ridge of water traveling on the surface of the sea; it is a manifestation of where the energy from the wind has been translated into circular movements of water molecules. The easiest way to understand this is to imagine a crowd performing a Mexican wave. The wave appears to move sideways around a sports stadium when, in fact, each person only moves up and down. In the case of ocean waves, the distance between adjacent wave crests is the wave length. The wave height is the vertical distance from the top of a crest to the bottom of a trough. When the wind blows gently, crests and troughs are smooth and round, producing a swell. When strong winds blow, the crests sharpen into peaks and the trough stretches out. The size of a wave depends upon the fetch—the distance over water that the wind blows in a given direction. Some of the biggest waves have been recorded in the Southern Ocean where, at some latitudes, the fetch runs right around the globe.

GLOBAL WAVE HEIGHT

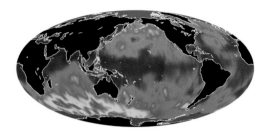

↑ **This map shows the global mean wave height.** Wave heights are color coded, from magenta for waves that are less than 3 feet (1 m), through blue, green and yellow to red for waves of 20 to 25 feet (7 to 8 m). The largest waves are found in the southern Indian Ocean and the Southern Ocean near Antarctica.

→ **When waves move** into shallow water, the circular motions in the water start to touch the seafloor, causing the waves to slow down. As the waves pile up close to the shore, the wavelength shortens and the wave becomes higher, the crest eventually toppling over and breaking up into surf.

← **Waves approaching** a steep shore will tend to rise up much more quickly than those moving over a gently sloping beach. This rapid rise in wave height means that the top of the wave is thin. The wave curls over and topples suddenly, creating an air-filled channel, or tube, between the crest and the foot of the wave.

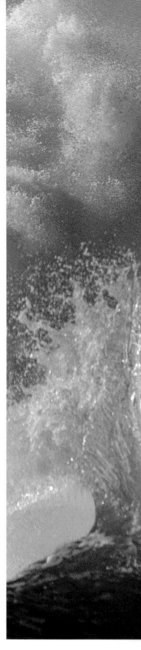

ROGUE WAVES

Winds do not blow uniformly, so the waves that they produce can interact in a number of ways. So-called rogue waves are created occasionally by several waves combining. It was once thought that there was a limit to wave height of 60 feet (18 m), and that reports of bigger waves were exaggerated. However, in 1933, a wave of 112 feet (34 m) was recorded by a reliable source. In 1996, the ocean liner *Queen Elizabeth II* was hit by a 95-foot (29 m) wave.

Movement of wind over the ocean generates corresponding movements in the water

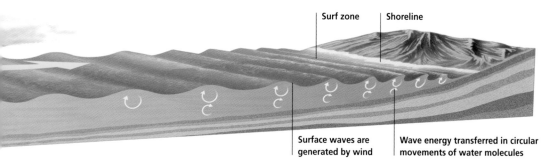

Surf zone | Shoreline

Surface waves are
generated by wind

Wave energy transferred in circular
movements of water molecules

WAVE GENERATION

The movement of winds over the oceans generates
corresponding movements in the water. Steady wind
over great distances tends to generate long, smooth
undulations of the surface (the swell). In the surf
zone close to the shore, water particles move in an
orbital motion and diminish with depth. When the
seafloor becomes shallow and the orbital motions
strike the bottom, they become flattened. Crests are
formed with the forward thrust, which leads to the
development of breakers.

Tsunamis

Tsunamis are sometimes incorrectly called tidal waves, but these highly destructive waves are not related to tidal activity. Their name is derived from a Japanese term meaning "harbor wave," because it is only when these waves reach shore that they manifest themselves and cause havoc. Tsunamis can be generated by earthquakes, undersea landslides, volcanic eruptions and large icebergs breaking away from glaciers. The most devastating tsunami activity of modern times occurred on December 26, 2004, following a massive undersea earthquake in the Indian Ocean—the most powerful since 1900. The tsunami destroyed coastal regions in Indonesia, Thailand, Sri Lanka, Myanmar, the Maldives and southern India; its effects were felt as far away as Africa. More than 290,000 people died as a direct result of the tsunami and millions more were displaced.

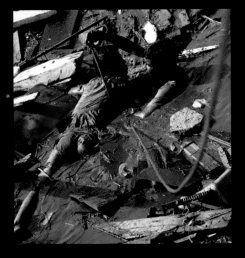

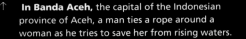

↑ **In Banda Aceh,** the capital of the Indonesian province of Aceh, a man ties a rope around a woman as he tries to save her from rising waters.

← **The satellite photograph at the top** shows part of the north coast of Aceh province, in Sumatra, Indonesia, in January 2003. Urban centers and lush green vegetation are clearly visible. Below is an image of the same area on December 29, 2004, three days after the tsunami hit. Vegetation has been stripped away and reclaimed land inundated. Indonesia was the country closest to the earthquake's epicenter.

TSUNAMI GENERATION

A sudden displacement of a large volume of water causes waves that spread in all directions from the initial disturbance—similar to the surrounding ripples when a stone is thrown into a pond. In the deep water of the open ocean, the waves generated are not hazardous. They are usually less than 40 inches (1 m) high and are barely noticed. However, once these waves reach shallow water, they rise up, producing the classic features of a tsunami. First, an unusually large amount of the sea pulls back from the shoreline, past the low watermark, to join with the developing tsunami. This is followed by one or more immense walls of water, up to 100 feet (30 m) high, that rush up the shore and push inland with unstoppable force.

↑ **Tsunami damage is not confined** to the shoreline, which takes the full force of the initial impact. The immense volume of water and the enormous speed of the wave give the tsunami sufficient momentum to carry seawater and debris far inland.

← *The Great Wave*, **by Japanese artist** Hokusai Katsushika, draws inspiration from the power and magnitude of tsunamis. On average, Japan experiences a tsunami every 6.7 years, the highest rate in the world.

Hurricanes are a spectacular and violent meteorological phenomenon caused by the interaction of the oceans and atmosphere. Known as cyclones in the Indo-Pacific region and typhoons in Japan, they form over warm, tropical waters, most often between 5 and 15 degrees latitude, slightly away from the equator. The warm water heats the air above, giving rise to clouds that become clusters of thunderstorms. The rotating storm system spins progressively quicker as it moves away from the equator. A storm must produce winds of over 74 miles per hour (119 km/h) to be classified as a hurricane. Once formed, it may last days or even weeks before sweeping poleward or crossing over land, often with disastrous results. Its severity and potential to do damage on land are rated on the five-point Saffir–Simpson scale (*see below*). Hurricanes frequently extend a great distance up into the atmosphere and can measure as much as 600 miles (970 km) across. Their formation and movement can be tracked by satellite.

THE SAFFIR–SIMPSON SCALE

	Pressure (hectopascals)	Wind speed (mph / km/h)	Storm surge (ft / m)	Damage type
1	more than 980	74–95 / 118–152	4–5 / 1.2–1.6	Minimal
2	965–979	96–110 / 153–176	6–8 / 1.7–2.5	Moderate
3	945–964	113–130 / 177–208	9–12 / 2.6–3.7	Extensive
4	920–944	131–155 / 209–248	13–18 / 3.8–5.4	Extreme
5	less than 920	more than 155 / 248	more than 18 / 5.4	Catastrophic

INSIDE A HURRICANE

A calm area of low pressure at the center, or eye, of a hurricane is surrounded by a spiraling wall of intense thunderstorms. The low pressure at the eye draws up mounds of water that will become a storm surge where the hurricane touches land. The strongest winds in a hurricane spiral outward—counterclockwise in the northern hemisphere and clockwise in the southern hemisphere.

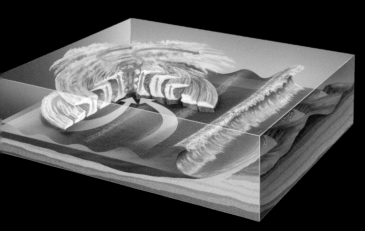

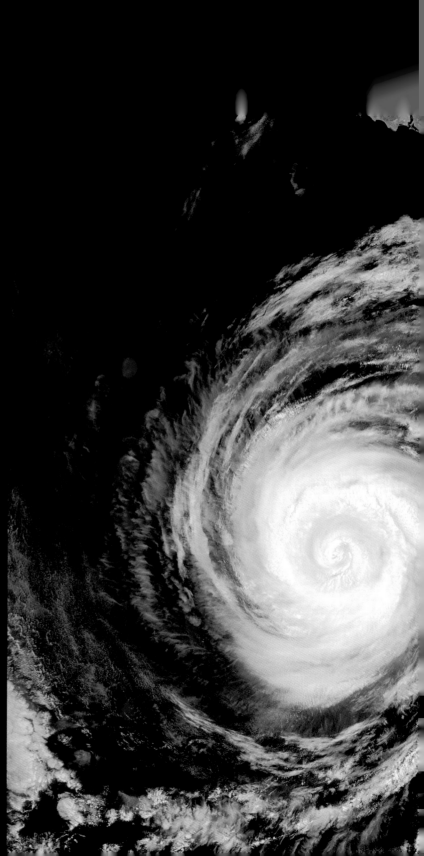

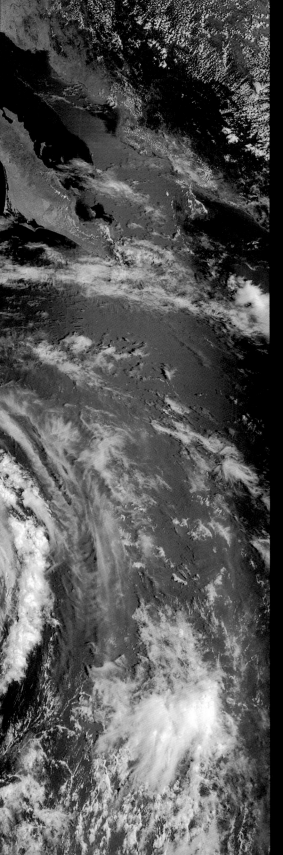

← **Hurricane Hernan off Baja California,** Mexico, in September 2002, peaked as a rare Category 5 hurricane, with winds of 155 miles per hour (248 km/h). It eventually moved away from the coast, and weakened to Category 3 hurricane status, with winds of 120 miles per hour (193 km/h).

→ **This color-enhanced** satellite image shows a hurricane approaching the coast of Bangladesh in 1991. The image is colored according to cloud density, from red (most dense), through yellow, green to blue (least dense). The most dense clouds are at the center, where the winds spiral rapidly outward from the eye.

← **Hurricane Isobel struck** North Carolina, USA, in September 2003, with devastating results. Hurricanes (and cyclones) are named alphabetically with alternating male and female names. Each basin or region has an agreed list of names but with different rules. For example, the first hurricanes in the Atlantic basin and Eastern Pacific of each year begin with an "A" name. The first cyclone of the year in the Central Pacific takes the next name on the list—regardless of its initial letter.

Ocean habitats

The annual progress of Earth around the Sun leads to the formation of distinct oceanic habitats. The parts of Earth receiving the most sunlight and heat throughout the year are the tropics, which lie between the Tropic of Cancer (23.5°N) and the Tropic of Capricorn (23.5°S), some 1597 miles (2570 km) north and south of the equator. Throughout the year, the Sun is directly over some latitude between the tropics. Since water temperatures are a key factor in shaping ocean environments, the tropics can also be defined as the zone in which surface water temperatures often reach 86°F (30°C) and rarely fall below 69°F (20°C).

The temperate zones are the intermediate zones that lie between the tropics and the polar regions. In the northern hemisphere, the temperate latitudes run from the Tropic of Cancer to the Arctic Circle (66.5°N), and in the southern hemisphere from the Tropic of Capricorn to the Antarctic Circle (66.5°S). The polar regions lie north and south of the Arctic and Antarctic circles.

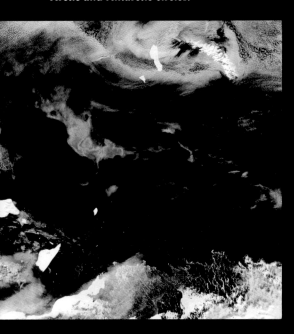

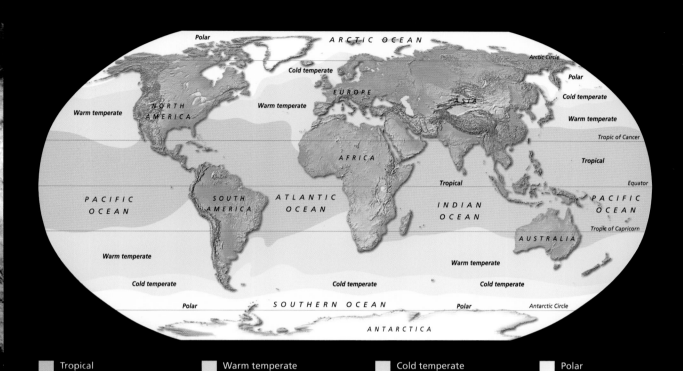

Legend:

Tropical
(over 69°F / 20°C)

Warm temperate
(50–69°F / 10–20°C)

Cold temperate
(40–50°F / 5–10°C)

Polar
(less than 40°F / 5°C)

↑ **Climatic zones based** on temperature do not coincide exactly with geographic zones because of the nature of ocean currents. For example, cold currents extend the warm temperate zones up the west coasts of Africa and South America.

← **Inshore tropical waters,** such as those around the Bahamas, are clear and blue due to a lack of phytoplankton production. The tropics do not have strongly defined seasons, with no major differences in the Sun's strength during the year, allowing phytoplankton to constantly reproduce at low levels.

←← **South Georgia lies** on the boundary between the polar and cold temperate zones, where the cold Antarctic water meets warmer, more saline, subantarctic surface water.

→ **In warm temperate zones,** such as the waters around New Zealand, there is a marked seasonality. Although the average water temperature is 50°F (10°C) and extremes of temperate are rare, the phytoplankton in the ocean have an intense spring growth phase known as the Spring Bloom.

Polar regions

The north and south polar regions contain the great ice sheets that cover 12 percent of the oceans. Both regions are cold because the tilt in Earth's axis limits the amount of solar radiation they receive. However, there is a distinct difference in the harshness of the weather between the two polar regions. In Antarctica, the coldest months produce temperatures ranging from -4° to -22°F (-20° to -30°C) on the coast and -40° to -94°F (-40° to -70°C) in the interior. Midsummer temperatures may reach as high as 59°F (15°C) on the Antarctic Peninsula, but the interior range is between -4° and -31°F (-20° and -35°C). By comparison, the Arctic is the warmer region, with mean temperatures there ranging from -31°F (-35°C) in the winter to 32°F (0°C) in the summer. There is concern that global warming is melting the polar ice. While evidence suggests that the Arctic ice sheet is indeed shrinking and thinning, there is disagreement among polar scientists about the predicted effects of warming on Antarctica.

ANTARCTIC ICE

ARCTIC ICE

Permanent sea ice
♦ ♦ ♦ Mean minimum ice limit
••• Mean maximum ice limit

POLAR ICE LIMITS

The Arctic ice sheet has a central portion that is permanently frozen. It covers around 2.7 million square miles (7 million km²). During the winter months, the ice sheet doubles to approximately 5.4 million square miles (14 million km²). In the Antarctic, there is a constant band of drifting sea ice around the continent, with the northern limit in the South Atlantic pack ice that extends to 52°S.

→ **The Drygalski ice tongue extends** from the David Glacier 50 miles (80 km) into the Ross Sea, Antarctica. Ice tongues occur when a glacier does not immediately break up into icebergs upon reaching water, but floats out to sea as a large mass. Once in the water, tidal movements help to break off icebergs from the ice tongue.

↑ **Most icebergs are calved** (created) by breaking off from the polar ice sheets. The Arctic produces the most icebergs—up to 15,000 per year. Fewer than 1000 per year are sighted further south than 48°N. By comparison, Antarctic icebergs have been seen as far away as Bermuda; this occurred in 1907 and 1926.

↓ **Isolation and harsh conditions** on Antarctica have severely limited its animal life to its shores and seas. Paradoxically, one of the main factors limiting life in polar regions is not the extreme cold, but the lack of liquid drinking water in regions that contain 75 percent of the world's freshwater.

Temperate regions

The temperate regions lie between the polar circles and the tropics in the northern and southern hemispheres. The regions have clearly defined seasons. This seasonality is the main factor contributing to the richness of temperate waters, ensuring that they have a high annual production of phytoplankton—the sudden upsurge in phytoplankton numbers is known as the Spring Bloom. In turn, the Spring Bloom fuels production of zooplankton that is transmitted up the food web to sustain higher grazers and predators such as fishes. Phytoplankton production is far greater in temperate regions than it is in tropical waters, where nutrients are limited. Polar waters are rich in nutrients, but ice cover and darkness severely restrict phytoplankton growth.

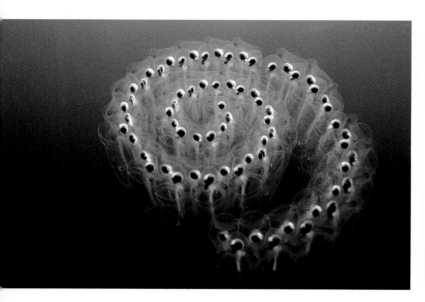

↑ **This salp and other phytoplankton grazers** benefit from the abundance of food in temperate waters. The presence of such large numbers of microscopic phytoplankton also affects the clarity and color of the water.

→ **The majority of the world's** commercially important finfish and shellfish species are caught in temperate waters. This abundance is dependent on the annual Spring Bloom, which initiates growth up the food web.

↗ **There are considerable regional** differences within the temperate zones. The waters of the Mediterranean are relatively clear, blue and unproductive compared to other temperate seas.

⇒ **Off the Californian coast,** the waters are highly productive because nutrients are brought to the surface by the mixing of coastal waters with the California Current. This current is also responsible for San Francisco's mild climate.

Tropical regions

Although light can penetrate deep into the clear waters of the tropics, the lack of defined seasons means that there is little mixing between the surface and deeper layers. Most of the nutrient-rich waters in the tropics lie below 492 feet (150 m), with the highest concentrations of nutrients found between 1640 and 3280 feet (500–1000 m). As a consequence, the growth of phytoplankton, which exists only in the surface layers of the oceans, is severely limited by a lack of phosphates and nitrates. The lack of nutrients in tropical surface waters accounts for the low concentrations of phytoplankton, which represent less than 1 percent of those found in temperate surface waters. However, there are exceptions to this lack of productivity. Equatorial upwellings bring surface-water nutrients to areas in the eastern Pacific, which in turn help to support the anchovy fisheries off Peru. Coastal upwellings also bring valuable amounts of nutrients to tropical coastal regions, such as the northwest coasts of Morocco and southwest Africa. Coral reefs are sites where nutrients slowly accumulate and are recycled. Coral polyps supply the nutrients that are lacking in the seawater, which increases the primary production by the coral's algal zooanthellae. The coral reefs are suitable sites for seaweeds to grow.

THE BLUE DESERT
Satellite images of the sea surface show that in the central tropical regions of the oceans, there is minimal primary production throughout the year. These areas are sometimes referred to as blue deserts, because the waters are a deep, clear blue but are lacking in marine life. Without the fuel of phytoplankton production, the numbers of small animals are very low and only large animals, such as blue marlin and albatrosses, are found crossing these regions.

↑ **The shallow, clear blue** waters and white coral sands of the Whitsunday Islands, within the Great Barrier Reef of Australia, are a major tourist attraction. The clarity of the water makes such areas particularly attractive to snorkelers and scuba divers, affording near-perfect visibility for studying reef life.

← **The abundant phytoplankton** that feeds the larger animals in temperate and polar food webs is absent in coral reef communities. Instead, animals feed on particles of organic material created by the breakdown of seaweeds and corals. Many small fishes and invertebrates also feed on the mucus produced by corals.

←← **Coral reefs are hot spots** of primary production in tropical waters and contain up to three times more plant than animal tissue. The zooanthellae in the coral account for only about 5 percent of a reef's biomass; the remaining biomass is mainly seaweed.

Exploring the oceans

Humans have long been fascinated by the sea and its riches, but the vastness of the oceans took centuries to explore and map. Today, the exploration of the oceans has moved far beneath the waves. Modern explorers have to meet technical challenges equivalent to space travel.

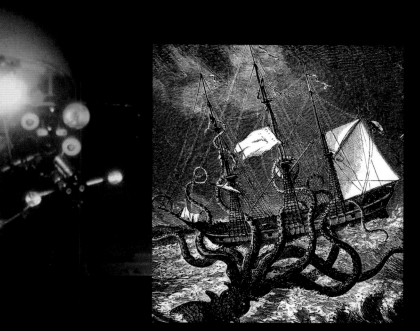

A history of ocean exploration

8000–7000 BC Dug-out wooden canoes in northwest Europe and reed boats in Mesopotamia and Egypt used to sail in rivers and coastal waters.

4000 BC Regular trade routes established on the Nile and waters around the delta.

4000–3000 BC First single-mast, square-rigged boats on the Nile.

2750 BC First recorded expedition of exploration from Egypt.

1200 BC The Phoenicians trade between the Mediterranean, the British Isles and West Africa in keeled boats with hulls built of planks. Boats carry tin, gold, spices and gemstones.

1000 BC Polynesians reach Tonga and Samoa.

325 BC The Greek navigator Pytheas sails to Britain and links lunar movements to the rise and fall of the tide.

c. 300 BC Chinese invent the compass, essential to voyages out of sight of land and at night.

AD 100 Chinese develop the rudder, making the development of large ocean-going vessels possible.

200 Chinese build the first multi-masted ships.

200–300 Development by Arabs and Romans of fore and aft rigging to allow boats to sail across the wind.

500 Polynesians cross the Pacific to colonize Hawaii.

780 Vikings cross the North Sea and raid the northwest coast of Europe.

850 Polynesians reach New Zealand.

1000 Eric the Red reaches Greenland from Iceland.

c. 1000 Lief Eriksson reaches North America via Greenland.

1400–1500 Three-masted ships with large hulls for cargo and supplies developed in northwest Europe.

1405–1433 Chinese explore the Indian Ocean and around the Cape of Good Hope.

1488 Portuguese navigator Bartolomeu Dias (1457–1500) rounds Cape of Good Hope.

1492 Genoese-born Christopher Columbus (1451–1506) makes first voyage to the Americas under the patronage of King Ferdinand and Queen Isabella of Spain.

1497 Portuguese navigator Vasco da Gama (1460–1542) reaches India by sea from Portugal.

1522 Portuguese navigator Ferdinand Magellan (1480–1521) completes the first circumnavigation of the world and names the Pacific Ocean.

1526 Sebastian Cabot (1474–1557), Venetian-born explorer and cartographer to Henry VIII of England, begins exploration of the Brazilian coast.

1606 Dutchman Willem Jantszoon (1571–1638) becomes the first European to sail to Australia.

1610 Englishman Henry Hudson (1565–1611) explores part of the Arctic Ocean and Hudson Bay.

1616 Dutchman Dirck Hartog (1580–unknown) charts the west coast of Australia.

1642 Abel Tasman (1603–1659) becomes the first European to sail to Tasmania, New Zealand, Tonga and the Fiji Islands.

1725 Danish explorer Vitus Bering (1681–1741) begins exploration of the seas off Alaska and northeastern Siberia and discovers sea route round Siberia to China.

1768 Englishman James Cook (1728–1779) makes a scientific voyage to Tahiti that includes mapping the coast of northern Australia.

1772 James Cook's second expedition takes him to Antarctica and Easter Island.

1774 Spanish navigator Juan Perez (1725–1775) explores the west coast of North America from Mexico up to Vancouver Island.

1777 First iron-hulled boat built in Yorkshire, England.

1783 Jouffroy d'Abbans (1751–1832) builds the first paddle-driven steamboat in France.

1786 French naval officer Jean-François La Perouse (1741–1788) maps the west coast of North America.

1795 Matthew Flinders (1774–1814) starts his circumnavigation and mapping of Australia.

1802 Scottish engineer William Symington (1763–1831) launches the *Charlotte Dundas*, the first stern paddle-wheel steamboat.

1807 The *Clermont* is the first commercial steamboat launched to carry freight and passengers between New York and Albany.

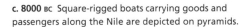

c. 8000 BC Square-rigged boats carrying goods and passengers along the Nile are depicted on pyramids.

1526 Sebastian Cabot, noted navigator, explores the Brazilian coast for the King of Spain.

1768 Englishman Captain James Cook begins his first voyage, reaching New Zealand and Australia in 1770.

1819 The *Savannah* makes the first steam crossing of the Atlantic.

1823 British naval officer James Weddell (1787–1834) discovers the Weddell Sea and the Weddell seal.

1831 Charles Darwin (1809–1882) departs on a five-year voyage aboard HMS *Beagle*.
British naval officer James Clark Ross locates the magnetic North Pole on June 1.

1836 British patent for the screw propeller granted to Francis Pettit-Smith (1808–1874).

1838 The *Great Western*, designed by British engineer Isambard Kingdom Brunel (1806–1859), begins first regular steamship crossings of Atlantic. Departure of the United States Exploring Expedition.

1841 James Clark Ross (1800–1862) charts much of the Antarctic coastline and discovers the Ross Sea.

1845 Isambard Kingdom Brunel's *Great Britain* is the first sea-going iron and propeller-driven ship to cross the Atlantic.
The *Rainbow*, the first clipper ship, launched in the United States.

1859 The *Great Eastern* launched. Designed by Isambard Kingdom Brunel, the ship combined paddlewheels with screw propulsion. It remained the largest ship built for nearly 50 years.

1863 British clippers *Taeping* and *Ariel* carry China tea to London in 99 days.

1869 Record-breaking clipper *Cutty Sark* launched. Originally built to carry tea from China, she carries wool from Australia to Britain in 78 days in 1883.

1872 Commencement of the first global oceanographic expedition on HMS *Challenger*.

1877 Start of the scientific cruises of the US Coast Survey vessel *Blake*, under the direction of Alexander Agassiz (1835–1910).

1886 The first boat powered by an internal combustion engine built by Gottlieb Daimler.

1893 Norwegian Fridtjof Nansen (1861–1930) reaches 86°N by drifting in Arctic iceflow in *Fram*.

1895–1898 Joshua Slocum (1844–1910) makes first single-handed circumnavigation of the world in his tiny sloop *Spray*.

1897 Canadian Charles Parsons (1854–1931) builds *Turbinia* to demonstrate power of marine steam turbines. Achieves 34 knots at Spithead naval review.

1900 Italian engineer Enrico Forlanini (1848–1930) builds the first hydrofoil.

1902 The first diesel-powered boat, the *Petit Pierre*, launched in France.

1910 *Selandia*, a Danish merchant ship, becomes the first ocean-going, diesel-powered ship.

1912 Sinking of RMS *Titanic* on maiden voyage after an iceberg collision, with loss of 1503 lives.

1925 Departure of United States *Meteor* expedition using the first echo sounder to map the seafloor.

1929 The British, Australian, New Zealand, Antarctic Research Expedition (BANZARE) begins the first complete mapping of the Antarctic under the leadership of Sir Douglas Mawson (1882–1958).

1947 Danish anthropologist Thor Heyerdahl (1914–2002) sets out on the *Kon-tiki* expedition, using a balsa wood raft, in an attempt to show the migration routes across the Pacific.

1950 Commencement of the Danish *Galathea* expedition to explore life in the deep ocean.

1955 First nuclear-powered submarine *Nautilus* launched in the USA.
British patent for the hovercraft granted to Christopher Cockerell (1910–1999).

1958 US nuclear submarine *Nautilus* makes the first submerged transit of the Arctic icepack passing through the North Pole.

1959 First nuclear-powered surface ships commissioned, the Soviet icebreaker *Lenin* and the US merchant ship *Savannah*.

1969–1970 Thor Heyerdahl's *Ra* expedition crosses the Atlantic in an ancient Egyptian-style papyrus reed boat.

1985 Location and inspection of the wreck of the *Titanic* by US oceanographer Robert Ballard, using remotely operated vehicle (ROV) *Argo*.

2004 Launch of *Queen Mary 2*—the largest passenger liner built to date.

1831 Charles Darwin begins a five-year voyage, collecting information used in his theory of evolution.

1912 *Titantic* sets sail from England on her maiden voyage, but is sunk by an iceberg off Newfoundland.

2004 The *Queen Mary 2* is launched. At 1132 feet (345 m) long, it is the world's largest passenger ship.

Ocean myths

The sea has been an integral part of the development of many civilizations, so it is no surprise that it has been associated with innumerable myths, legends and deities still recognized today. The most comprehensive cultural associations with the sea are those of the various island groups of Oceania. There, most aspects of the sea's attributes have a specific deity or legend associated with them. Other peoples have imbued the sea, and many of its plants and animals, with mystical and cultural significance. For example, the indigenous people of the Pacific northwest of America still carve totem poles to represent the essence and bounty of the oceans. In western culture, the vast Greek and Roman pantheon of sea gods and legends is indicative of the importance of the sea to these civilizations.

→ **This painting, *Angelica saved by Ruggiero*,** by Ingres (1780–1867), shows Ruggiero, riding on a hippogriff (a mythological animal, half horse and half griffin), rescuing Angelica from a sea monster.

↓ **In Greek mythology,** Earth was thought to be a flat disk encircled by an unending stream of water—an early concept of the ocean. This is a representation of Oceanus, the god who personified this stream.

MODERN MYTH

The Bermuda Triangle—an area of the tropical Atlantic between Bermuda, Florida and Puerto Rico—has become a modern myth of the sea. It is supposedly the site of innumerable mysterious disappearances of ships and aircraft. Many explanations have been offered to account for noted disappearances. However, comparison with other sea areas, with similar weather conditions and traffic, shows that disappearances are no greater than might be expected.

The Kraken, a legendary sea monster in Scandinavian mythology, rose from the depths to attack ships. It is possible that the legend was based on sightings of giant squid.

The sirens were daughters of the sea god Phorcys. They lured sailors to their doom with their beautiful singing. Odysseus outwitted them by blocking his crew's ears with wax. He then lashed himself to the mast in order to hear the sirens' voices. *The Sirens*, by Edouard Veith, was painted in 1895.

Early voyages

The earliest records of sea voyages are from ancient Egypt, dating back 5000 years. At about the same time, a number of other seafaring civilizations developed around the shores of the Mediterranean and the Middle East. There are records of long sea voyages and information on navigation from Phoenician, Greek, Persian, Roman and Arabic sources. By 1250 BC, the Phoenicians had ventured beyond the Mediterranean and were trading with Britain and West Africa. In the Pacific, Polynesian peoples began their spread across the island chains around AD 500. In northern Europe, the Vikings were the first to explore beyond the coastal waters of Scandinavia, reaching North America around AD 1000. From around AD 200 onward, the Indian Ocean was explored by Arab traders and Chinese trading fleets.

↖ **One of the earliest** surviving world maps is on a Babylonian clay tablet, dating from about 600 BC. Babylon is shown at the center, and the world is surrounded by one ocean.

↓ **This ancient Egyptian relief** depicts a boat with passengers. Around 3000 BC, development of single-masted, square-rigged vessels greatly increased the speed and possible size of ships.

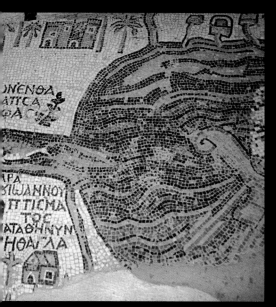

↑ **This ancient Greek vase** depicts a sea battle. Seamanship and navigation were essential in ancient Greece. City states were reliant on maritime trade routes and used fleets of fast galleys to protect cargo vessels and to deter invaders.

→ **This finely carved** ship is a replica of a Viking craft found in Norway in 1904. Around AD 1000, the Vikings frequently made long sea journeys. Voyages in replicas of Viking ships, using the authentic Viking navigation methods, proved that this type of craft could survive the conditions of the open ocean and was able to cross the Atlantic.

↑ **This sixth-century AD mosaic** depicts the Dead Sea, Jordan River and surrounding towns. Early explorers had to rely on maps that owed much more to imagination than geographical accuracy.

Traversing the oceans

Humans have always been driven by a desire to find out what lies over the horizon. Today, no blank spaces are left on the world map. However, it has taken millennia of exploration to piece the map together. The first records of voyages of exploration describe the journeys of an Egyptian explorer who traveled to Arabia in 2750 BC. In the Pacific, Polynesians set out on their long voyages using outrigger canoes and astronomical navigation to reach Tonga and Samoa in 1000 BC. European, Native American, Chinese and Muslim explorers and navigators all spread out from their homelands so that the world's oceans were gradually charted. As navigational knowledge and instrumentation improved, voyages of exploration became longer. By the eighteenth century, only the inhospitable seas of Antarctica remained uncharted.

↓ **Between 1578 and 1580** Sir Francis Drake circumnavigated the world in the *Golden Hinde*, a replica of which is pictured here. The ship was only 120 feet (37 m) long but carried an 80-man crew.

← **Captain James Cook** possessed exceptional skills in seamanship and navigation. He undertook three major voyages of exploration between 1769 and 1779, and collected valuable scientific information.

→ **The sextant was developed** in 1731 from older instruments such as the octant. The sextant helped to make accurate navigation possible.

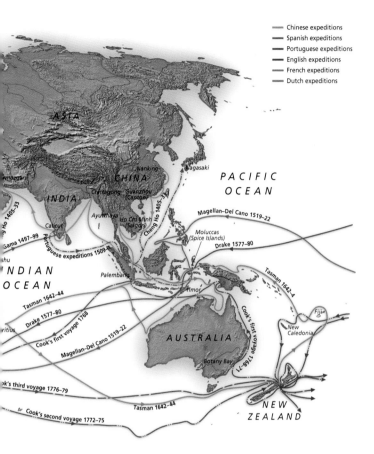

Chinese expeditions
Spanish expeditions
Portuguese expeditions
English expeditions
French expeditions
Dutch expeditions

AMERICAN WORLD

The young empires of the Americas had little time to expand before the arrival of Europeans. They were also restricted by geography: the Aztecs had deserts to the north and jungle to the south, while the Incas were wedged between the Andes and the Pacific.

THE AMERICAN WORLD
Extent of Aztec empire
World known to Aztecs c.1500
Extent of Inca empire
World known to Incas c.1500

EUROPEAN WORLD

By the late 15th century, European traders and travelers had compiled a detailed knowledge of what was to them the Old World, from the west coast of Africa to the East Indies. Columbus's voyage of 1492 would add the Americas to their maps.

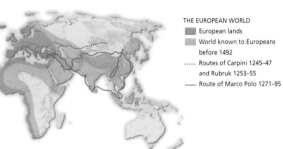

THE EUROPEAN WORLD
European lands
World known to Europeans before 1492
...... Routes of Carpini 1245–47 and Rubruk 1253–55
—— Route of Marco Polo 1271–95

ISLAMIC WORLD

In 1500, the culture of Islam was much more extensive than that of Europe's Christendom. The world knowledge of Muslims had been substantially augmented by the intrepid Ibn Buttuta, who visited India, China, Sri Lanka, Sumatra and Africa.

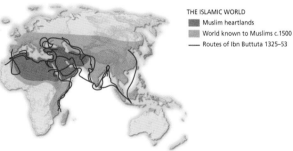

THE ISLAMIC WORLD
Muslim heartlands
World known to Muslims c.1500
—— Routes of Ibn Buttuta 1325–53

ASIAN WORLD

In the 15th century, China had the world's largest population and most advanced technology. In spite of isolationist policies, it had amassed substantial world knowledge and continued to do so with the sea voyages of the Ming admiral Cheng Ho.

THE ASIAN WORLD
Territory of China, Japan, Korea
World known to the Chinese c.1500
...... Route of Cheng Ho 1405–33
—— Silk Road trading route

Reaching Antarctica

The first mention of Antarctic exploration dates back to the seventh century, when Maori oral histories describe the voyage of a Polynesian war canoe that sailed south and reached a frozen ocean. Medieval scholars and navigators speculated on the existence of an unknown southern continent, matching the frozen wastes of the Arctic, but it was not until the second voyage of James Cook that the existence of Antarctica was considered a reality. During the nineteenth and early twentieth centuries, the major European powers vied with one another to explore Antarctica and reach the South Pole. Today Antarctica is protected by an international treaty that prohibits any activity likely to harm this last wilderness. Antarctica is studied by scientists from several nations that carry out joint studies of this unique environment.

→ **The only exploitation** of Antarctica permitted by the 1961 treaty is the growing activity of ecotourism. Passenger ships now regularly visit the southern continent.

↘ **The US Antarctic expedition** of 1928. The following year, four of the crew members carried out the first flight over the South Pole.

→ **During his second voyage** of 1772–75, James Cook sailed round the globe in the high southern latitudes and ventured into pack ice between 60° and 70°S, crossing the Antarctic Circle three times. He did not sight Antarctica, but information he brought back encouraged others to venture south.

ANTARCTIC EXPLORATION TIMELINE	
Year	Event
1773	Englishman James Cook crosses the Antarctic Circle and circumnavigates Antarctica.
1819–21	Russian Fabian von Bellingshausen circumnavigates the Antarctic and discovers some offshore islands.
1820	First sighting of Antarctica by British naval officers, William Smith and Edward Bransfield.
1821	First landing on continental Antarctica by American sealer Captain John Davis. Disputed.
1823	British whaler James Weddell discovers the sea named for him and discovers the most southerly point at that time.
1840s	Separate British, French and American expeditions establish the status of Antarctica as a continent after sailing along the coastline.
1899	Carsten Borchgrevink leads a British expedition landing at Cape Adare. Believed to be the first confirmed landing.
1909	Australian Douglas Mawson reaches the South Magnetic Pole.
1911	Norwegian Roald Amundsen reaches the South Pole.

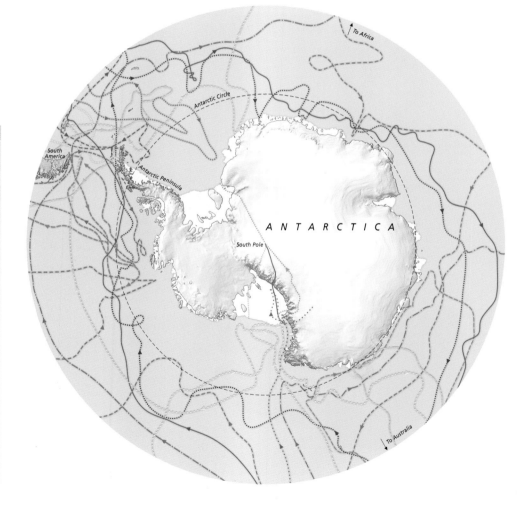

The modern era

The replacement of sail by steam, and wood by iron and then steel, ushered in a new era of cheap and reliable sea transport. Technological innovations and the building of ever-larger ships were driven by commercial rivalry between shipping companies to capture the most lucrative routes. The competition was fiercest on the North Atlantic crossing, where the numbers of ships and frequency of sailings were such that people referred to the "Atlantic Ferry." The golden age of the great transatlantic liners reached its high point in the 1930s, when the French *Normandie* and the British *Queen Mary* tried to surpass each other in speed and comfort. However, the competition from airlines in the 1950s and 1960s made liners uneconomical, resulting in most of them being scrapped. A few ships began to carry passengers on warm-water cruises and, by the mid-1980s, the popularity of cruising saw a revival of the fortunes of large passenger liners. This revival of the ocean liners culminated in the 2004 launch of the *Queen Mary 2*, the largest passenger vessel built to date.

THE BLUE RIBAND
The Blue Riband is a hypothetical honor awarded to a ship making the fastest transatlantic crossing between the lights off the Scilly Isles and the entrance to New York harbor. The last liner to hold the honor was the *United States*. She made the crossing in 1952 in 3 days, 10 hours and 31 minutes.

↓ **The Cunard liner *Queen Mary 2*** made her maiden voyage from Southampton, England, in January 2004, carrying a full complement of 2600 passengers. *Queen Mary 2* is the largest passenger vessel launched; she is 1138 feet (347 m) long and 238 feet (72.5 m) high.

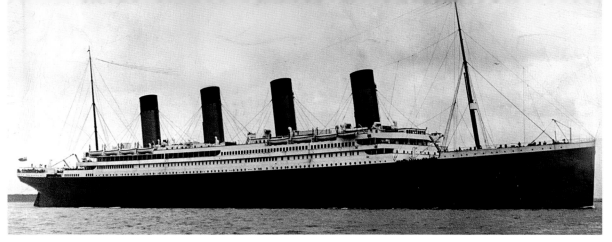

← **The ill-fated *Titanic*** was the second of three sister ships built for the White Star line as a means of breaking the Cunard line's hold on the transatlantic route. Competition was based on speed and comfort.

← **Until the launch of *Queen Mary 2*,** the *Star Princess*, operated by P&O Cruises, was described as the world's largest moving object, carrying the same number of passengers as *Queen Mary 2*. The demand for cruising holidays continues to grow, stimulating the building of ever-larger ships with facilities such as golf courses, shopping malls and even planetariums.

← ***The World* is a new variant** on traditional cruising that set out on her first world voyage in 2002. The ship's accommodation consists of private apartments that are rented on a time-share basis for short-duration cruises, or purchased by customers who can take up permanent residence.

Marine archaeology

Although scattered artefacts from shipwrecks and submerged settlements have been washed ashore or salvaged for centuries, the methodical archaeological processing of underwater sites is a relatively recent development. Marine archaeology owes its existence to the advent of scuba diving after the Second World War. The first significant underwater excavation was in 1960, when a 3000-year-old ship was discovered off the Mediterranean coast of Turkey, still with its cargo of wine jars. Since then, many important finds have been made, including the discovery and retrieval of historic ships such as Viking longships in Denmark, the sixteenth-century English warship *Mary Rose*, and the *Vasa* raised from Stockholm harbor. Not all marine archaeological sites are shipwrecks. Fragments of Pharos lighthouse, one of the seven wonders of the ancient world, were found during recent excavations of the sunken port of Alexandria in Egypt.

↓ **The remains of a boat** that sank off eastern Turkey in 1025 are mapped with the use of a grid. Cargo and other items can then go to the surface. The Mediterranean is rich in marine archaeological sites because of the long history of marine commerce and the violence of the area's winter storms.

The *Vasa*, a Swedish warship, sank on her maiden voyage in 1628, and was raised in 1961. The cold waters of the Baltic with their low salinity, and the thick mud of Stockholm harbor inhibited the work of wood-boring animals and decay organisms; most of the ship is in near-perfect condition.

5000 years of oceanography

3800 BC First known maps showing Egypt's waterways.

2300 BC First documented explorer, Egyptian nobleman Harkuf, leads expeditions up the Nile to the land of Yam (southern Nubia).

900 BC Greeks first use the word "okeanes" to describe the sea. This is the root of the modern word ocean.

800 BC First navigational charts produced.

230 BC Eratosthenes makes the first estimate of Earth's circumference and invents the concept of latitude and longitude.

200 BC The astrolabe invented—a tool to measure angles between the Sun, horizon and principal stars. The earliest form was used by Greek sailors; refined over centuries, it was not displaced as an essential navigational tool until the 18th century.

127 BC Hipparchus (190–120 BC) creates a regular grid of latitude and longitude to improve navigation.

AD 1000 Vikings use a combination of celestial observations, mainly of the midday Sun, using a Sun compass and the Pole Star at night to navigate for long periods out of sight of land.

1419 Prince Henry of Portugal (1394–1460) sets up the first school for navigators.

1537 Flemish mapmaker Geradus Mercator (1518–1594) designs a map that is specifically for navigation. The Mercator map was the first successful means of representing a spherical shape of Earth on a flat map, rather than a globe.

The method of representation, known as the projection, distorts the shapes of objects and distances, moving away from the equator. Many other cartographers have devised projections to overcome this problem but Mercator's projection is still one of the most familiar.

1610 Galileo Galilei (1564–1642) invents the telescope.

1675 King Charles II (1630–1685) establishes the Royal Observatory at Greenwich to produce astronomical tables for navigation, particularly to solve the problem of determining longitude at sea. Latitude could be determined by simple observations of the Pole Star.

1707 The loss of four English warships and the death of 2000 men, including Admiral Sir Cloudsley Shovell (1650–1707), wrecked on the Scilly Isles due to inaccuracies in determining longitude.

1714 The Board of Longitude set up in England. It offers a £20,000 prize to anyone with a practical means of determining longitude at sea. Longitude could be determined only by extremely complex astronomical comparisons. The need was for a means of accurately measuring time at sea, but the only accurate clocks were based on the pendulum, which would not work on a moving ship. The self-taught English clockmaker John Harrison (1693–1776) spent the next 50 years developing an accurate chronometer in order to claim the prize.

1730 The modern sextant invented by John Hadley (1682–1744). The Hadley sextant replaced the variety of astrolabes and other navigational instruments used. Today, it still continues to be a valuable navigational instrument despite the introduction of satellite navigation systems.

1760 John Harrison presents his H4 chronometer to the Board of Longitude. This provides a simple means of determining longitude at sea.

1769 Benjamin Franklin (1706–1790) publishes the first chart to show an ocean current (the Gulf Stream).

1818 John Ross (1777–1856) collects the first deep-water sediment samples.

1835 Gaspard Coriolis publishes his treatise on the influence of Earth's rotation on moving objects, leading to an understanding of winds and currents.

1836 William Harvey produces the first coherent classification of seaweeds.

1843 Edward Forbes presents his Azoic hypothesis, based on his deep-water dredging expeditions. He concludes the abundance of marine life diminishes with depth and that the floor of the oceans would be completely devoid of life. Although disputed at the time and later disproved, the hypothesis did stimulate exploration of the deep seas.

Liverpool tidal observatory, now the Proudman Oceanographic Laboratory, set up to formulate tidal predictions for mariners around the world.

127 BC Hipparchus employs simple sighting devices to chart stars, using them to establish position at sea.

1610 Galileo is the first person to make a telescope that could be used for astronomy and navigation.

1730 The development of an accurate sextant is made possible by precision machine tools.

1855 Matthew Fontaine Maury (1806–1873) publishes the *Physical Geography of the Seas*—the first comprehensive work on physical oceanography.

1865 Pietro Secchi (1818–1878) uses a plate lowered into water to measure turbidity. The disk is still in use for rapid assessment of water conditions.

1868 Thomas Huxley (1825–1895) proposes that material found in deep-sea sediment samples is living protoplasm that covers the deep ocean floor and may be the origin of life. He names the material *Bathybius huxleyi*. It is later shown to be an artefact created when salts in seawater are mixed with alcohol preservative.

1880 William Dittmar identifies and measures the major salts in seawater.

1884 The Marine Biological Association of the United Kingdom founded.

1888 The Marine Biological Laboratory founded at Woods Hole, Massachusetts, USA.

1891 John Murray and Alphonse Renard produce the first classification of marine sediments.

1902 The International Council for the Exploration of the Seas founded in Copenhagen.

1905 First regular wireless time signals broadcast from Washington, D.C. for use by mariners.

1910 Prince Albert I of Monaco founds the Monaco Oceanographic Museum and Laboratory dedicated to marine science.

1912 Alfred Wegener presents his continental drift hypothesis in Frankfurt.
The International Iceberg Patrol set up as a direct consequence of the loss of the *Titanic*.
Scripps Institution for Biological Research joins the University of California and becomes the Scripps Institution of Oceanography.

1919 The British Discovery committee set up to organize Antarctic research and oceanography in the Southern Ocean.

1921 The International Hydrographic Bureau founded. The Soviet Marine Research Institute, later the Shirshov Institute of Oceanology, set up in Moscow.

1925 The German Meteor expedition begins the first comprehensive mapping of plankton. The expedition also tests the first echo-sounder system.

1930 Woods Hole Oceanographic Institution founded.

1949 The National Institute of Oceanography (NIO) formed in the United Kingdom. It later becomes part of the Southampton Oceanography Centre. Lamont-Doherty Geological Laboratory established, specializing in underwater geophysical observations.

1960 Harry Hess and Robert Dietz propose that the ocean basins are formed by spreading from the mid-ocean ridges and that crustal material is lost at ocean-margin subduction zones.

1962 The Ocean Research Institute incorporated into the University of Tokyo.

1965 The work of John Tuzo Wilson and others leads to the concept of plate tectonics and confirmation of Wegener's continental drift hypothesis.

1968 The *Glomar Challenger* drilling ship provides evidence to support the plate tectonic concept. Bedford Institute of Oceanography opens in Canada.

1970 National Oceanic and Atmospheric Administration (NOAA) set up in the USA.

1974 The Australian Institute of Marine Science (AIMS) opens in Townsville, Queensland.

1977 The US deep-diving submersible *Alvin* finds hydrothermal vents and previously unknown marine communities in the Galapagos rift zone.

1978 Launch of *Seasat*—the first satellite dedicated to oceanographic measurements.
Launch of Global Positioning System (GPS), operated by the US to give continuous worldwide coverage.

1980 The First Institute of Oceanography (FIO) established in Qindao, China.

1984 French Research Institute for Exploitation of the Sea (IFREMER) founded in France.

1992 The TOPEX-Poseidon satellite launched to improve oceanographic observations from space.

1998 The British unmanned submersible *Autosub* begins its first mission to make continuous oceanographic measurements in the world's oceans.
The *Galileo* spacecraft finds evidence of an ocean under the surface of one of Jupiter's moons, Europa.

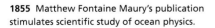

1855 Matthew Fontaine Maury's publication stimulates scientific study of ocean physics.

1910 The Monaco Oceanographic Museum and Laboratory is founded.

1987 *Nautile* is launched, operated by IFREMER. It is capable of diving to 20,000 feet (6000 m).

Major centers of oceanography

The importance of the oceans in human affairs can be judged by the numerous major centers of oceanographic research and study throughout the world. Another indicator of the interest in understanding the oceans is that governments continue to invest in new centers, equipped with the latest research facilities and vessels. The oceans hold many secrets yet to be discovered.

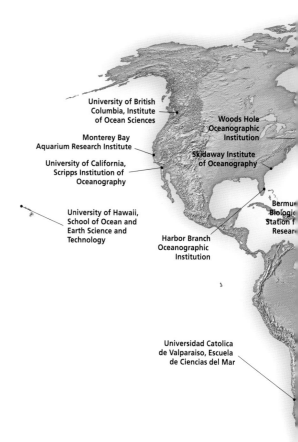

THE AMERICAS AND AFRICA

Bermuda Biological Station for Research (Bermuda)
Dalhousie University,
Department of Oceanography (Canada)
Dartmouth Bedford Institute of Oceanography,
Ocean Sciences Division (Canada)
Harbor Branch Oceanographic Institution (Florida)
Monterey Bay Aquarium Research Institute (California)
Skidaway Institute of Oceanography (Georgia)
St John Memorial University of Newfoundland
Ocean Sciences Center (Canada)
Universidad Catolica de Valparaiso,
Escuela de Ciencias del Mar (Chile)
University of British Columbia,
Institute of Ocean Sciences (Canada)
University of California,
Scripps Institution of Oceanography (California)
University of Cape Town,
Center for Marine Studies (South Africa)
University of Hawaii, School of Ocean and Earth
Science and Technology (Hawaii)
University of São Paulo,
Oceanographic Institution (Brazil)
Woods Hole Oceanographic Institution (Massachusetts)

ASIA AND AUSTRALASIA

Australian Institute of Marine Science (Australia)
CSIRO Marine Laboratories (Australia)
Japan Marine Science and Technology Center (Japan)
National Institute of Oceanography (India)
Ocean University of Qingdao (China)
University of Hong Kong,
Swire Institute of Marine Science (Hong Kong)
University of Otago,
Department of Marine Science (New Zealand)
University of Tokyo, Ocean Research Institute (Japan)

EUROPE

Alfred Wegener Institute for Polar and
Marine Research (Germany)
French Research Institute for Exploitation
of the Sea (France)
International Hydrographic Bureau (Monaco)
Netherlands Institute for Sea Research (Netherlands)
Shirshov Institute of Oceanology (Russia)
Southampton Oceanography Centre (United Kingdom)
Université de Bretagne Occidentale, Institut
Universitaire Européen de la Mer (France)
University of Kiel, Institute of Marine Research (Germany)

The Australian Institute of Marine Science (AIMS) was established in 1972 to research the ocean environment.

The Alfred Wegener Institute focuses on research in the polar regions and their surrounding seas.

Woods Hole Oceanographic Institution is a group of laboratories that work on all aspects of oceanography.

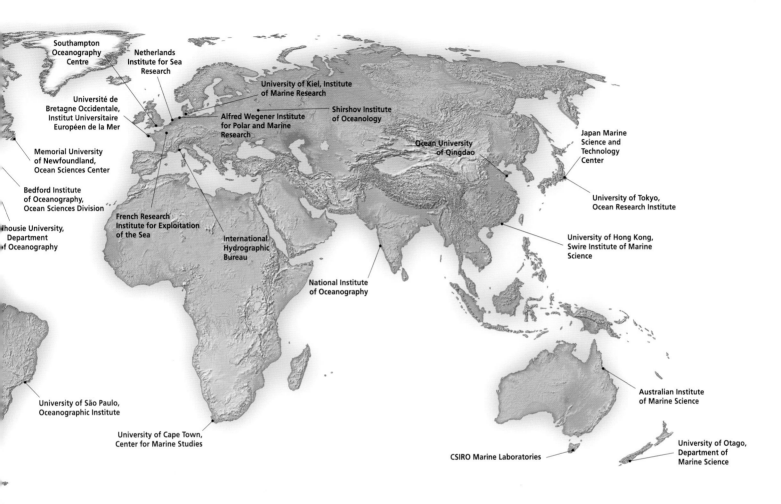

Southampton
Oceanography
Centre

Netherlands
Institute for Sea
Research

University of Kiel, Institute
of Marine Research

Université de
Bretagne Occidentale,
Institut Universitaire
Européen de la Mer

Alfred Wegener Institute
for Polar and Marine
Research

Shirshov Institute
of Oceanology

Memorial University
of Newfoundland,
Ocean Sciences Center

Ocean University
of Qingdao

Japan Marine
Science and
Technology
Center

Bedford Institute
of Oceanography,
Ocean Sciences Division

University of Tokyo,
Ocean Research Institute

lhousie University,
Department
f Oceanography

French Research
Institute for Exploitation
of the Sea

International
Hydrographic
Bureau

University of Hong Kong,
Swire Institute of Marine
Science

National Institute
of Oceanography

University of São Paulo,
Oceanographic Institute

Australian Institute
of Marine Science

University of Cape Town,
Center for Marine Studies

University of Otago,
Department of
Marine Science

CSIRO Marine Laboratories

Harbor Branch Oceanographic Institution conducts research from the top of the shore to the deep sea.

Southampton Oceanography Centre was opened in 1995. It conducts university teaching and research.

Japan Marine Science and Technology Center conducts research on extreme environments and climate change.

Our knowledge of the basic geography of the oceans has taken many centuries to accumulate. This was due to the technical problems of position fixing and obtaining information from an alien environment into which humans could not venture. From the eighteenth century onward, with the birth of modern navigation, the shapes of the oceans at the surface could be mapped, often with a degree of accuracy that was not surpassed until the advent of aircraft and satellite imaging of Earth's surface from the 1970s onward. Today, the science of remote sensing—that is, using airborne or satellite cameras and other sensors—not only provides high-resolution images but also accesses a wealth of other chemical and physical details about the oceans, such as temperature, chlorophyll, the presence of oil slicks and wave heights. Mapping the topography of the seafloor and its underlying geology on a global scale was revolutionized by the development of seismic and sonar systems that provided the first detailed information of the structure of the ocean basins in the 1960s.

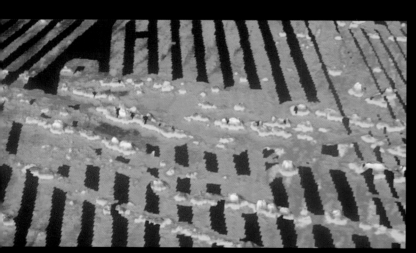

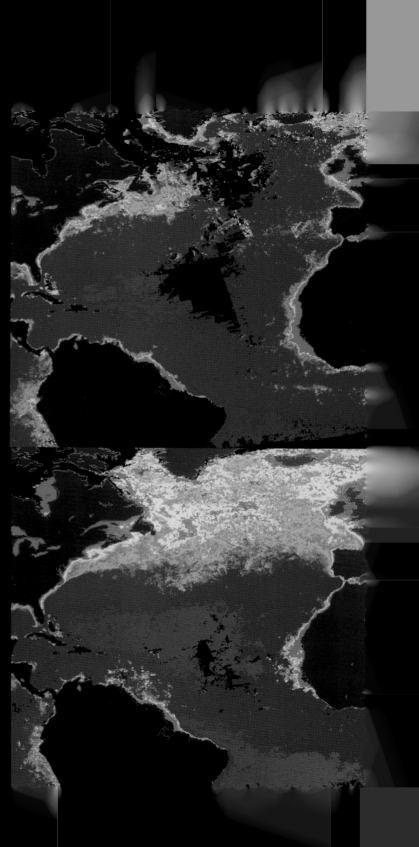

↑ **With modern navigational** aids, survey ships can tow sonar devices in regular patterns over wide areas of the ocean floor. The colors on the image indicate the height of the features on the ocean floor, with blue the lowest and white the highest points. The image shows volcanoes up to 6560 feet (2000 m) high on the floor of the east Pacific

→ **Satellite imaging can follow** ocean processes changing with time—something not possible with traditional ship-based sampling methods. Here, satellite measurements show development of surface chlorophyll concentrations in winter (above) and during the North Atlantic Spring Bloom (below).

SURVEYING WITH SOUND

Side-scan sonar uses beams of sound waves that are sent out from a towed emitter. Changes in the sound that is reflected back are processed to give visual representations of the seafloor. Some side-scan sonars can penetrate into the upper layers of the seabed, showing its structure. However, to map the deeper strata of the ocean floor, it is necessary to use seismic surveying. Intense bursts of sound are produced by powerful underwater electrical discharges (known as sparkers), compressed air guns or small explosive charges. The sound that is then reflected back is detected by hydrophone arrays towed behind the ship.

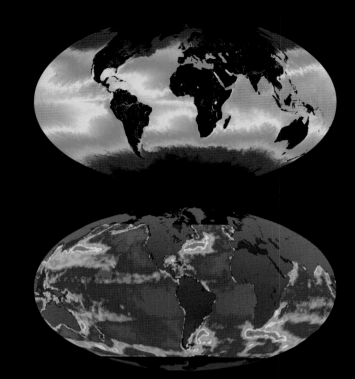

← **Satellite-borne infrared** sensors can build up a global image (*above*) of the distribution of heat in the oceans by detecting heat emitted from the sea surface. The orange and yellows are the warmest parts and the blues and greens are the coldest. Paths of ocean currents (*below*) can be traced using remote sensing, showing their tracks through the oceans and how they can change in size and direction over time. Currents can be detected using sensors to measure small changes in sea surface height (radar altimetry) that show up as red patches.

SEAFLOOR MAPPING

1. Visual and seismic survey vessel

2. Intense sound source

3. Strata beneath seabed

4. Sound waves

5. Reflected sound waves

6. Cable depressor

7. Camera, lights and other sensors

8. Drogue

9. Acoustic signals

10. Side-scan emitter

11. Hydrophone array

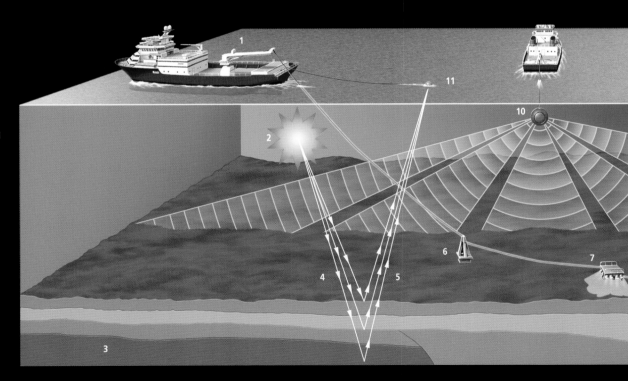

Sampling the water column

Oceanographers studying the water column have to find ways to sample a vast three-dimensional medium and deal with samples and processes that occur on microscopic to global scales. Traditional sampling methods tend to be snapshots of one small spot, at one instant in time. Continuous, real-time sensors have overcome some of these challenges. Advances in underwater photography and acoustic imaging have revolutionized the sampling of ocean life, although nets and water bottles are still used to verify information from new technology.

An array of water bottles is lowered to predetermined depths. The stoppers at either end are closed by a signal sent from the ship. This allows samples from a precise point in the water column to be retrieved and analyzed.

The size range of plankton is such that it cannot be sampled with a single net. "Bongo" plankton nets are pairs of nets of two mesh sizes that can sample plankton over a wider size range in the water column.

CURRENT METER

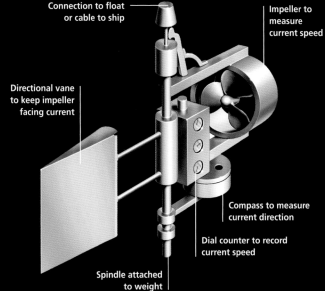

Connection to float or cable to ship

Impeller to measure current speed

Directional vane to keep impeller facing current

Compass to measure current direction

Dial counter to record current speed

Spindle attached to weight

SAMPLING WITH SOUND

Sonar was first used in the 1930s to show features on the seabed. Since then, scientists have considered using sonar beams to follow the physical processes in the water column, such as currents that can be picked out where differences in temperature or salinity make them stand out acoustically from the surrounding water. However, the technology to do this effectively did not become available until the early 1980s with the development of the Acoustic Doppler Current Profiler (ADCP). It was quickly realized that the ADCP could also be used to follow plankton movements. Today, it is possible to count numbers of organisms and estimate their sizes with the use of sonar.

← **Measurement of water** movements can be done either by following drifters on the surface, or by using simple current meters such as the one shown here.

↓ **Plankton nets are towed slowly** through the water, either horizontally or vertically, so that water is funneled into the net and organisms filtered out.

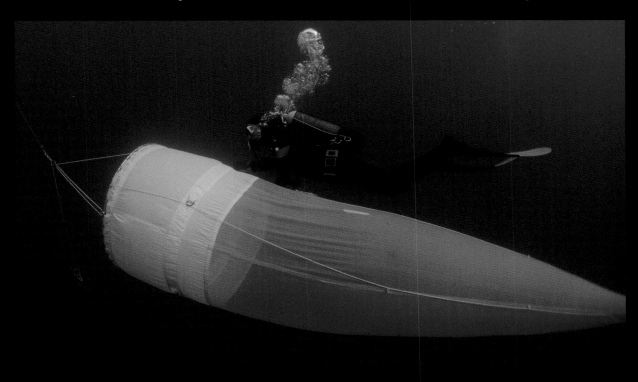

Sampling the seafloor

Sampling the seafloor has been likened to trying to deduce the complexities of a modern city by collecting samples from a balloon, the ground obscured by cloud. The scale of the task is also illustrated by the estimate that, since the voyage of HMS *Challenger* more than 130 years ago, less than 4 square miles (10 km²) of the 139.4 million square miles (361 million km²) of the ocean floor have been sampled scientifically. Only one-tenth of this sampling effort has been fully analyzed and published in scientific journals. Sampling the seabed can be divided into two main forms: sampling on the surface and covering a known area; or taking a defined sample by digging into the upper layers of the sediments that cover most of the ocean floor. Sampling gear has also been developed to ensure samples are retrieved in the best possible condition, without the loss of samples or contamination from material in the water column.

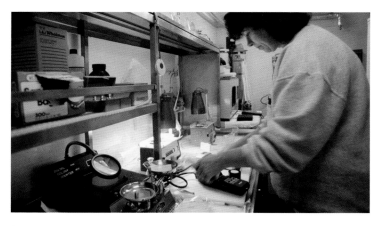

↑ **Once samples are retrieved,** they are analyzed or stored for further study. Many research vessels have specialized containerized laboratories that can be changed according to the research to be carried out on each voyage.

Trawls
Trawling will show what animals are present, but it is not possible to sample an accurately defined area of seabed and trawl samples are described as qualitative, rather than quantitative. **1.** The Agassiz trawl is used in coastal waters and can be up to 10 feet (3 m) wide. **2.** The mouth of the beam trawl is held open in a frame up to 20 feet (6 m) across, but the difficulty of handling long beams at sea limits its size. **3.** The semi-balloon otter trawl can be much wider—up to 48 feet (14.5 m) because it has no beam. Instead, the otter boards at the sides of the mouth pull open the net when it is towed underwater.

Epibenthic sled
Surface-living animals too small to be trawled are sampled using an epibenthic sled. Hinged plates close the mouths of the nets on the journey to and from the surface and an odometer wheel records the distance traveled on the seafloor.

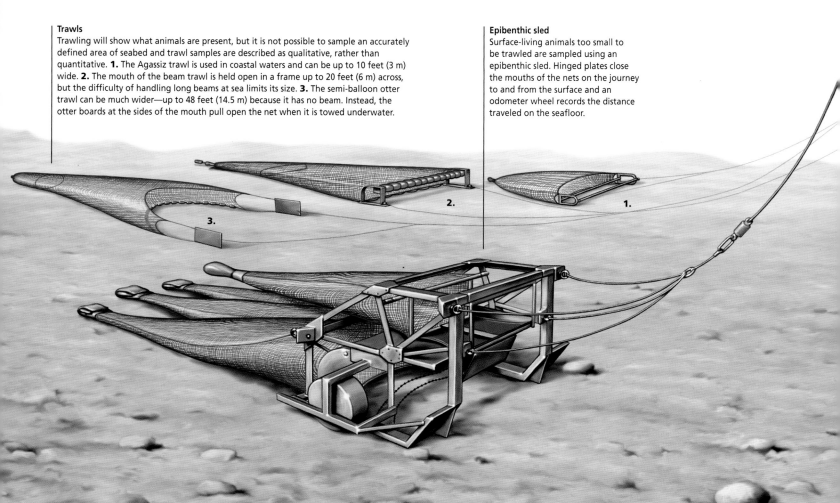

Spade box-corer
A spade box-corer collects single samples 20 x 20 inches (50 x 50 cm). The frame sits on the surface while the corer is pushed into sediment. Upon removal, a spade is drawn under the corer to seal it.

Piston corer
Long cores are obtained with a piston corer. This is a heavily weighted tube, dropped from above the seabed when a trigger weight touches bottom. Pulling the main cable pulls up a piston inside the tube, forcing the corer deeper, before retrieval.

Multi-corer
The multi-corer takes cores of sediment with a known total surface area. Once on the bottom, a set of open plastic tubes is slowly pushed into the seabed. As the tubes are pulled out, the ends are automatically capped.

Anchor box dredge
As it is drawn along, an anchor box dredge gathers surface sediments until it is full. A hinged plate at the front protects the sample on the way up. The sample is then retrieved from a door at the rear of the dredge.

A history of submersibles

500 BC Herodotus gives an account of diving used in warfare by Greek diver Scyllis, who cuts loose the fleet of Persian King Xerxes I off Cape Artemesium.

AD 1538 First written account of a diving bell in Spain.

1620 Dutchman Cornelius Drebbel demonstrates his 12-oared, wood and leather submarine to James I of England. The boat uses tubes attached to floats to carry air from the surface, presaging the development of the "schnorkel" during World War II.

1650 Otto Von Guericke develops the first air pump.

1667 Robert Boyle (1627–1691) records observation of decompression sickness or "the Bends" when he observes gas bubbles in the eye of a viper that had been compressed and then decompressed.

1690 Edmund Halley (1656–1742) patents a diving bell, connected by pipe to weighted barrels of air replenished from the surface. The device was of practical use and manned bells were lowered to depths of 65 feet (20 m) for up to 1½ hours.

1715 John Lethbridge builds a "diving engine," a submarine and diving suit hybrid, constructed from an oak barrel and supplied with air from surface pumps. It was claimed a diver could stay submerged for 30 minutes at a depth of 60 feet (18 m).

1776 US engineer David Bushnell builds the hand-powered, one-man submarine *Turtle*, which makes use of buoyancy tanks, as a weapon of war to attack the British fleet in New York harbor.

1790 Improved version of Halley's diving bell by US inventor John Smeaton (1724–1794) is used for salvage work in Ramsgate Harbor, England.

1823 English inventor Charles Anthony Deane patents a "smoke helmet," supplied with air through a hose, for fighting fires. The value of the helmet in diving is quickly recognized.

1825 The elements of a practical Self-Contained Underwater Breathing Apparatus (SCUBA) are put together by Englishman William James. The diver carries a reservoir of compressed air on a belt fastened around the body.

1837 German-born English inventor Augustus Siebe (1788–1872) attaches Charles Deane's diving helmet with a watertight seal to a waterproof rubberized canvas suit. The closed suit and helmet are supplied with air from the surface and are considered to mark the start of modern diving.

1839 Siebe's "Improved Diving Dress" is used during the salvage of the British warship HMS *Royal George* that sank in 65 feet (20 m) of water at Spithead in 1783. Accounts of the salvage, which continued until 1843, indicate that the divers suffered symptoms of "rheumatism and cold," that were likely to have been the first recorded symptoms of decompression sickness in humans.

1843 Following the salvage of HMS *Royal George*, the Royal Navy sets up the world's first diving school.

1855 German Wilhelm Bauer, builds the submarine *Le Diable-Marin* for the Russian navy. It is hand-cranked and makes 134 dives before sinking.

1863 Launch of *Le Plongeur* designed by Bourgois and Brun to overcome the problem of propulsion. At 141 feet (43 m) long, *Le Plongeur* was mainly occupied by air tanks to power a compressed air engine; the design was also innovative in using the compressed air supply to empty her ballast tanks.

1864 The CSS *Hunley* becomes the first submarine to sink a surface ship, the USS *Housatonic* in Charleston Harbor, USA. *Hunley* was modified from a ship's boiler, powered by eight men cranking a propeller and armed with a spar torpedo.

1865 French engineer Benoit Rouquayrol and naval officer Auguste Denayrouse patent the "Aerophore" for underwater breathing. It consists of a horizontal steel tank of compressed air strapped to the diver's back and connected to a mouthpiece through a regulating valve. The valve was innovative in two ways. Firstly, it only let air through when the diver breathed in. Secondly, it was sensitive to the external water pressure, adjusting the delivery pressure according to depth—the same principle that operates in modern diving demand valves. The system was limited by the short endurance of the tank and normally used with a hose connected to a surface pump via the reservoir tank.

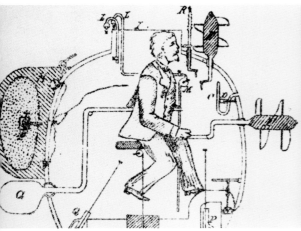

1776 Bushnell's *Turtle* is a one-man submarine that uses an underwater propeller for the first time.

1839 "Hard-hat" divers have to work restricted by an airhose connected to a pump on the surface.

1864 The USS *Housatonic* is the first warship to be sunk by a submarine, the CSS *Hunley*.

1873 Decompression sickness, also known as "Caisson Disease" or "the Bends," is formally described by surgeon Dr Andrew H. Smith He was asked to investigate the problem during construction of the Brooklyn Bridge in New York. Although Smith did not know the cause to be the development of nitrogen bubbles in the blood vessels, joints and spinal nerves, his report recommended that recompression chambers should be used to treat workers showing symptoms.

1876 Englishman Henry A. Fleuss develops the first practical, self-contained diving apparatus, capable of sustained use independent of a surface supply. Fleuss's equipment uses compressed oxygen rather than air and is a forerunner of modern closed-circuit scuba units used by military divers. The system was capable of sustaining a diver for periods up to 3 hours but only at shallow depths, since pure oxygen becomes extremely toxic below about 26 feet (8 m)—a fact then unknown.

1878 French physiologist Paul Bert publishes *La Pression Barometrique*, which showed that decompression sickness is due to the formation of nitrogen gas bubbles. Bert also suggests that gradual ascent to allow the safe dispersion of the nitrogen would avoid the problem, and that the condition could be successfully treated by recompression and then slow decompression.

1880 Englishman Reverend George Garrett successfully demonstrates a steam-driven submarine, the *Resurgam*. Steam was first generated in a coal-fired boiler on the surface. The craft then retained sufficient heat to drive underwater for some time.

1885 French submarine *Goubet I* is launched. Her electric motor system marks the start of practical underwater propulsion systems.

1897 Irish–American John Philip Holland (1840–1914) designs the *Holland IV*—the prototype of the modern submarine—combining the internal combustion engine for surface use and charging lead–acid accumulators that power an electric motor when underwater.

1908 Scottish physiologist John Scott Haldane (1860–1936) publishes a landmark paper on the prevention of decompression sickness with Arthur E. Boycott and Guybon C. Damant, providing the first scientific rules for staged decompression. These were immediately adopted by the Royal Navy and other navies around the world.

1924 Nitrogen narcosis or "rapture of the deep" caused by deep diving on compressed air leads to the first experimental helium-oxygen dives in trials conducted by the US Navy and Bureau of Mines.

1930 William Beebe and Otis Barton descend 1427 feet (435 m) off Bermuda in a steel sphere, or bathyscaphe, only 5 feet (1.5 m) in diameter.

1933 Rubber fins for underwater swimming patented by Frenchman Louis de Corlieu.

1935 The French navy adopts the SCUBA equipment modified from the Rouquayrol-Denayrouse system by naval officer Yves Le Prieur. He redesigns the demand valve to work with much higher pressures, which greatly increases the duration of diving independent of a surface supply.

1942–43 Jacques-Yves Cousteau (1910–1997), a French naval lieutenant, and Emile Gagnan, a gas engineer, redesign the demand valve. Its simple and rugged construction made mass production possible, bringing scuba diving into the realms of the affordable for a wide range of scientific, commercial and recreational users.

1960 Swiss engineer Jacques Piccard and US navy lieutenant Don Walsh descend 35, 814 feet (10,916 m) to the lowest point in the oceans—the Marianas Trench—in the bathyscaphe *Trieste*.

1965 The Woods Hole Oceanographic Institution (WHOI) launches *Alvin,* the renowned submersible.

1981 Launch of the *Shinkai 6500* in Japan—the deepest-diving manned submersible in commission.

1989 *Shinkai 6500* reaches 21,409 feet (6527 m). This is the deepest dive by a manned submersible in current use.

1995 *Keiko* sets a new depth record for a ROV at 36,008 feet (10,978 m) in the *Challenger Deep*.

1873 Working at high air pressure on the Brooklyn Bridge causes decompression sickness in laborers.

1897 John Philip Holland is the first to design a prototype of a modern submarine.

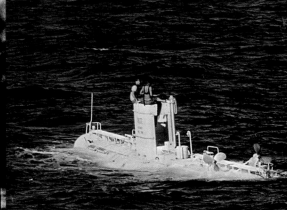

1960 Piccard and Walsh emerge from the hatch of the bathyscaphe *Trieste* after their record deep dive.

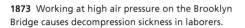

Manned submersibles

Manned submersibles are small submarines used to explore the oceans. Their design and construction are such that they can descend to greater depths than conventional submarines and carry out complex underwater tasks. Scientific exploration of the ocean depths by manned craft commenced in 1934, when Americans William Beebe and Otis Barton descended to 3024 feet (922 m) in a bathyscapthe—a heavy, hollow steel sphere suspended from the surface by a thin cable that carried a telephone line and power for a single electric light bulb. The world record for a manned dive is still held by the bathyscapthe *Trieste*; it descended to 35,815 feet (10,916 m) in the Mariana Trench in 1960. Since then, manned submersibles have undergone extensive developments. They were first equipped for military purposes, but are now used for scientific research, underwater surveying, construction and maintenance. The latest generation of manned submersibles can reach and explore most deep trench systems.

→ **Sea Cliff was originally** built for the US Navy, but since 1998, has carried out scientific exploration for the Woods Hole Oceanographic Institution.

→→ **The three-man crew** of *Sea Cliff* sits inside a hollow titanium sphere. Crews working on submersibles must cope with many hours in confined conditions.

↘ **MIR I and MIR II** are a pair of Russian manned submersibles.These submersibles undertake both scientific research and underwater tourist trips.

SUBMERSIBLES IN SERVICE		
Submersible	Launch (modified)	Depth feet (m)
Alvin	1964 (1973)	15,000 (4500)
Argos	1975	2000 (600)
Clelia	1974	1150 (3500)
Cyana	1970	10,000 (3000)
Deep Rover	1984	3300 (1000)
Deep Rover	1994	3300 (1000)
Deep Rover II	1994	3300 (1000)
Delta	1982	1000 (300)
Diving Saucer	1959	1150 (350)
DSRV 1	1971	3500 (1000)
DSRV 2	1976	5000 (1500)
Jago	1990	1650 (500)
Johnson Sea Link 1	1971	2600 (800)
Johnson Sea Link 2	1975	2600 (800)
MIR I	1987	20,000 (6000)
MIR II	1987	20,000 (6000)
Nautille	1985	20,000 (6000)
Nekton Gamm	1971	1000 (300)
NR-1	1969	2300 (700)
OSMOTR	1985	1000 (300)
Pisces IV	1971	6500 (2000)
Pisces V	1973	5000 (1500)
Pisces VII	1975	6500 (2000)
Pisces XI	1975	6500 (2000)
Saga	1987	2000 (600)
Sea Cliff	1964 (1982)	20,000 (6000)
Shinkai 2000	1981	6500 (2000)
Shinkai 6500	1987	21,400 (6500)
SM80/2	1990	1650 (500)
Snooper	1969	1000 (300)
SO-450-Vaimana	1982	1500 (450)
Turtle	1968 (1985)	10,000 (3000)

↓ **Nautile is operated** by IFREMER, the French marine research organization. This submersible can dive to extreme depths, and can explore 97 percent of the ocean floor. Its dive record includes visiting the wreck of the *Titanic*.

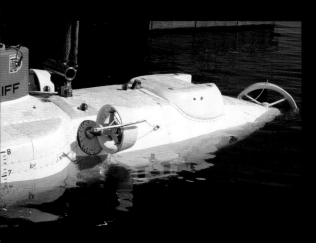

Alvin

Alvin is the most well-known manned submersible still in service. The Woods Hole Oceanographic Institution (WHOI) has operated the Deep Submergence Vehicle (DSV) since its launch in 1965. The vehicle is owned by the United States navy. In 1973, *Alvin* underwent major modifications that increased her maximum depth range to 14,764 feet (4500 m). A typical dive takes two scientists and a pilot on an eight-hour mission—two hours in descent and ascent, and four hours on the seafloor.

During her long career, *Alvin* has been associated with many important missions in underwater science and exploration. The submersible first made headlines in 1966, when she successfully recovered a hydrogen bomb that had been accidently dropped into the depths of the Mediterranean Sea. In 1974, *Alvin* took part in the project that explored the Atlantic mid-ocean ridge, and produced the first detailed images and samples which helped to confirm the nature of seafloor spreading. During a similar research program around the Galapagos rift zone in 1977, *Alvin* discovered hydrothermal vents and the animal life they supported. Here was the first known example of animals sustained by chemical energy rather than sunlight. Similar communities were found around cold water seeps during *Alvin* dives in 1984, in the Gulf of Mexico. In 1986, *Alvin* was one of the first submersibles to survey the sunken *Titanic*.

DEEP SEA SANDWICHES

An accident in 1968 gave an unexpected insight in to the microbiology of the deep sea. When support cables failed, *Alvin* sank, hatches open, to 5000 feet (1524 m). Although unmanned, the submersible contained a packed lunch in an open leather satchel. The satchel protected the food from animals but not from micro-organisms. When *Alvin* was recovered 10 months later, the meat sandwiches and apple were in a remarkable state of preservation, indicating the slow rates of microbial decay in the deep sea.

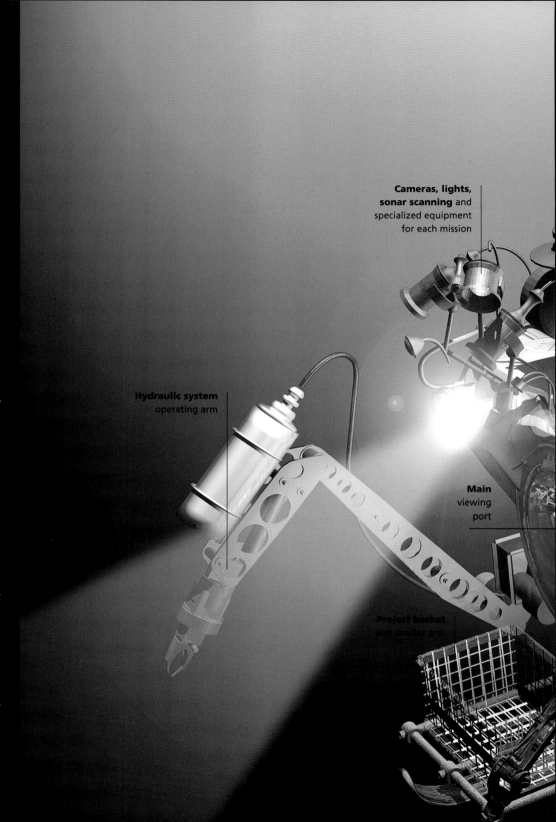

Cameras, lights, sonar scanning and specialized equipment for each mission

Hydraulic system operating arm

Main viewing port

Project basket and leather arm

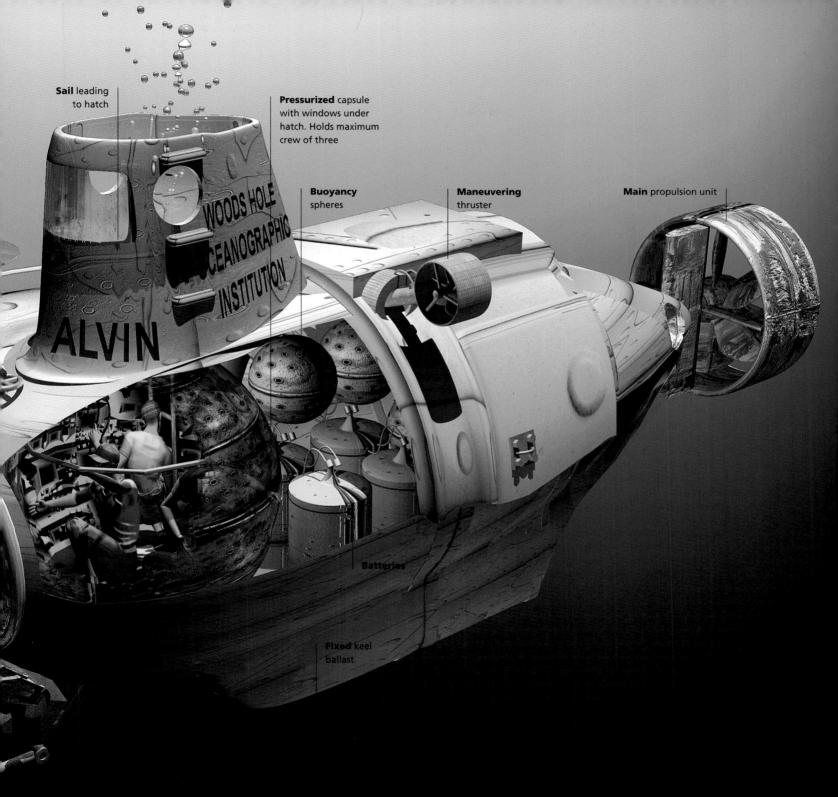

Sail leading to hatch

Pressurized capsule with windows under hatch. Holds maximum crew of three

Buoyancy spheres

Maneuvering thruster

Main propulsion unit

WOODS HOLE OCEANOGRAPHIC INSTITUTION

ALVIN

Batteries

Fixed keel ballast

Unmanned submersibles

Unmanned submersibles are the third generation of marine research vessels, after surface ships and manned submersibles. The enormous strides made in underwater robotics, photography and onboard guidance systems in the last 20 years have transformed unmanned submersibles. From simple camera frames, these submersibles have developed into vehicles that can augment or replace manned vehicles, removing the risks and limitations inherent in sending humans into the ocean depths. There are two types of vehicle in use for scientific, military and commercial uses. Remotely operated vehicles (ROVs) are controlled by an umbilical cable carrying signals and power from the surface and relaying back information. Autonomous underwater vehicles (AUVs) carry their own power supply and guidance system, enabling them to run on preset missions. AUVs relay information by acoustic links or by surfacing periodically to transmit stored data via satellite.

EXPLORING SPACE

Lake Vostock is a huge lake 2½ miles (4 km) below the Antarctic ice sheet. Its waters are kept from freezing by geothermal energy generated by the hot rocks of the continental crust below the lake. It is thought that the conditions in the lake are likely to be similar to those beneath the surface of Europa, the ice-covered moon of Jupiter. The AUV technology and experience gained from sampling Lake Vostock is likely to be utilized when a probe is sent to inspect Europa.

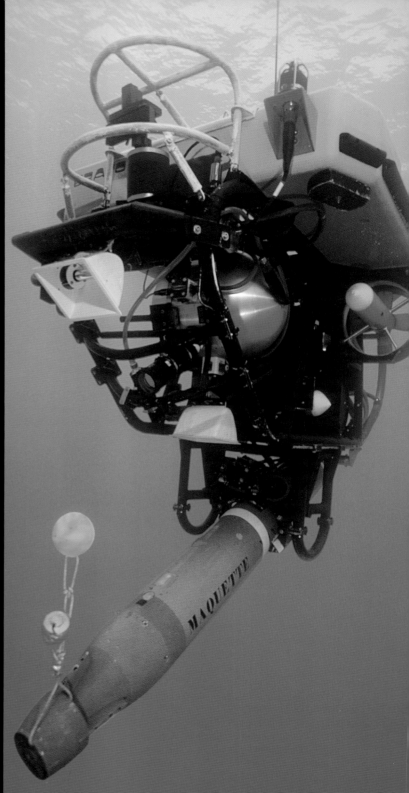

↑ **Increasingly sophisticated** robotic systems are being attached to ROVs and manned submersibles. Arms, such as these, must be able to function under the temperature and

→ **The ROV *Erato*** is a recovery vehicle developed by the French navy. The vehicle is fitted with specialized sensors to augment visual identification. It is seen

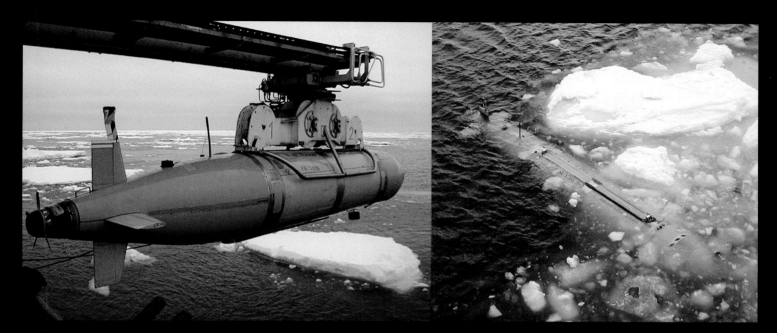

↑
↗ **Autosub is an AUV.** Its missions are operated by the Southampton Oceanography Centre. Guided only by an on-board computer, this AUV has been used to take measurements in the water beneath the Antarctic pack ice. This would be impossible to do from surface ships and dangerous for a manned submersible. AUVs are not restricted by an umbilical cable, such as those attached to ROVs.

→ **ANGUS is an early-model** camera sled ROV, renowned for surveying the Galapagos Rift in 1977 and for detecting hot water from the hydrothermal vents. The vents were investigated later in *Alvin* dives.

Early diving

The mystery of what lay hidden beneath the sea was long a subject of conjecture. Direct contact with the underwater world was limited to the duration of human breath. Although pearl and sponge divers could work on the seabed for a longer time than most people, mechanical means were needed to make it possible to stay underwater for prolonged periods—to collect its riches, salvage lost ships and cargo, and to sabotage enemy ships in wartime. It was not until the mid-seventeenth century that the invention of the first successful diving apparatus—the diving bell—made the dream of working underwater a practical reality. Later developments were designed to allow divers to remain underwater for longer periods and to move with greater freedom.

→ **1690 Halley** Earliest diving bell. Maximum depth 60 feet (18 m).

1715 Lethbridge Armored diving suit. Maximum depth 72 feet (22 m).

1828 Deane Earliest diving helmet. Maximum depth 80 feet (24 m).

1865 Rouquayrol and Denayrouze Development of the demand valve, which allowed surface air to be used by the diver. Maximum depth 100 feet (30 m).

1878 Fleuss Earliest self-contained breathing apparatus. Maximum depth 60 feet (18 m).

1918 Ohgushi Peerless respirator supplied air at the correct pressure through the diver's inflatable belt. Maximum depth 300 feet (91 m).

1943 Cousteau and Gagnan Development of the aqualung. Maximum depth 200 feet (61 m).

← **This underwater photograph** from 1893 gives some impression of the difficult conditions experienced by early divers.

1865 Rouquay and Denayrou demand valve

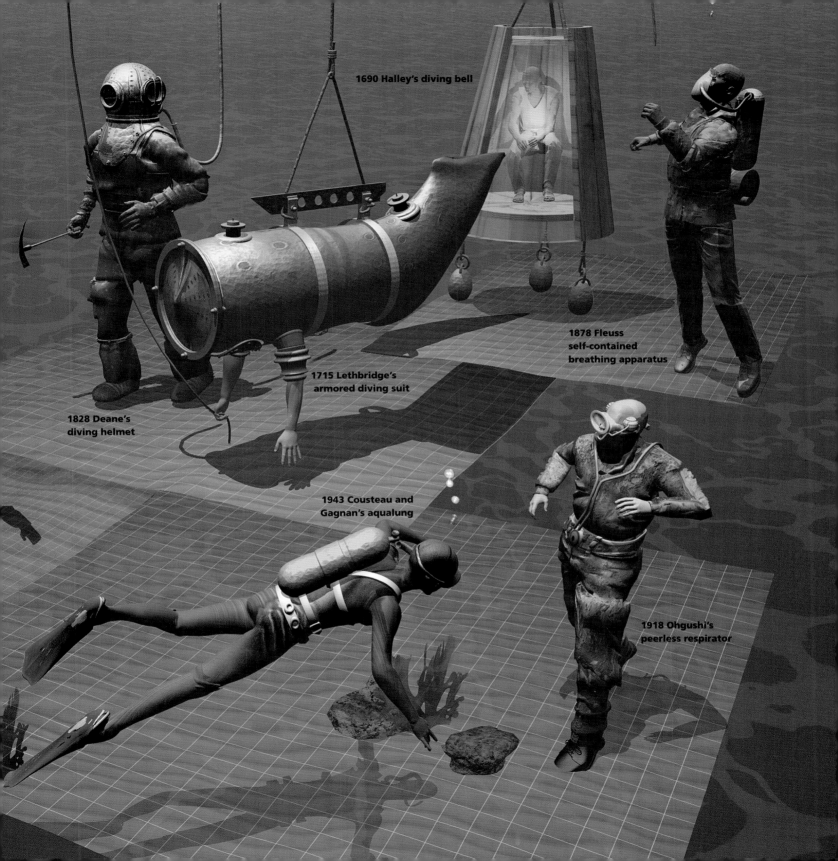

1690 Halley's diving bell

1878 Fleuss self-contained breathing apparatus

1715 Lethbridge's armored diving suit

1828 Deane's diving helmet

1943 Cousteau and Gagnan's aqualung

1918 Ohgushi's peerless respirator

Modern diving

The development of modern diving techniques has pursued two separate paths. The rediscovery of the demand valve in the 1940s, and development of lightweight high-pressure air cylinders, freed divers from clumsy helmets and air lines. These changes transformed diving from a difficult and expensive work task into a recreational activity enjoyed by millions of people around the world. Scuba diving has also made it possible for marine biologists and archaeologists to carry out underwater studies that would have once been too costly or difficult. To enable divers to work at depths that are not accessible to scuba divers, armored, atmospheric suits have been developed. These suits protect divers from external water pressure so that they can breathe air normally—at the same pressure as that on land. Decompression stops on the ascent are not needed.

↓ **The Newtsuit is a diving suit** that makes possible a descent of 2000 feet (600 m). Its design enables the diver to breathe uncompressed air. A thruster backpack is used to move around and the flexible joints allow great dexterity.

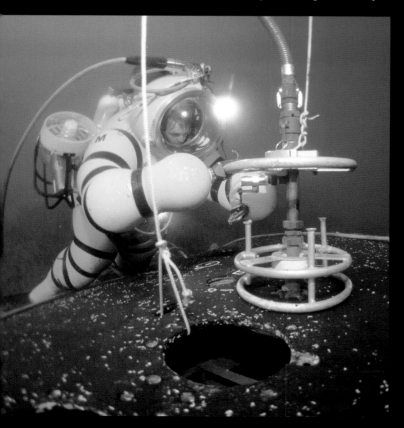

MODERN SCUBA

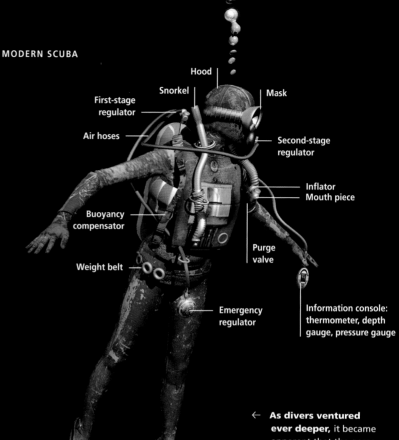

- Hood
- Snorkel
- Mask
- First-stage regulator
- Air hoses
- Second-stage regulator
- Inflator
- Mouth piece
- Buoyancy compensator
- Purge valve
- Weight belt
- Emergency regulator
- Information console: thermometer, depth gauge, pressure gauge
- Fin blade

← **As divers ventured ever deeper,** it became apparent that the gases in the air we breathe become toxic under pressure. For example, nitrogen starts to produce a narcotic effect from depths below 50 feet (15 m). To allow divers to go deeper, gas mixtures have been formulated to replace the nitrogen and reduce the oxygen content.

→ **The ability to explore the diversity** and beauty of underwater life has been completely transformed by modern scuba-diving equipment. It has enabled people to view marine life in its natural surroundings, and has generated great interest in the oceans.

DIVING METHOD AND DEPTH	
Diving method	**Maximum depth limit**
Diving bell (1690)	60 feet (18 m)
Earliest scuba (1878)	60 feet (18 m)
Armored diving suit (1715)	72 feet (22 m)
Diving helmet (1828)	80 feet (24 m)
Back air tank (1865)	100 feet (30 m)
Aqualung (1943)	200 feet (60 m)
Peerless respirator (1918)	300 feet (90 m)
Free diving (2003)	400 feet (122 m)
Scuba, air only	475 feet (145 m)
Helium–oxygen mix	2000 feet (600 m)
Newtsuit armored suit (1985)	2000 feet (600 m)
Helium–oxygen–nitrogen mix	2300 feet (700 m)

Ocean life

Life in the ocean ranges from the smallest bacteria to the blue whale, the largest living creature that has ever lived on Earth. A survey of the groups of marine plants and animals reveals an abundance of species, with unique adaptations to their oceanic habitats.

The food web

In all ecosystems there is a continual flow of energy and matter between the organisms that inhabit a given area. This transfer of energy comprises a systematic series of steps that can be traced by following feeding, or trophic, relationships. Organisms can be divided into the primary producers that make food from inorganic molecules, and the consumers that feed on them.

At the base of any food web, all but a few ecosystems depend on plants to trap solar energy and produce new plant material from nutrients. In marine ecosystems, the bulk of solar energy is taken up by phytoplankton, with a small contribution from seaweeds, sea-grass beds and mangroves. In turn, this plant material is either grazed by zooplankton and larger filter feeders, or dies and falls to the bottom to provide food for organisms on the seabed. Zooplankton, in their turn, are grazed by larger animals and, like the phytoplankton, when they die they also carry organic matter down to the seabed. Material and energy are also taken out of the marine food web by seabirds.

The quantity of material that flows up the food web decreases at every level; this limits the numbers of top-level consumers that the ecosystem can sustain. In their turn, these animals return material to the bottom of the food web through their waste products and decayed matter when they die.

THE TROPHIC PYRAMID
In the food web (*right*), energy and materials flow from a broad base of billions of small producers through a decreasing number of larger organisms to a few top consumers—this is known as the trophic pyramid. Studies suggest this structure occurs because only 10 percent of the energy entering each level is passed up to the next stage. Energy and materials at the top of the pyramid are, therefore, a fraction of those represented by the primary producers.

Solar energy drives the food web

Plants trap the energy of the Sun (primary producers)

Phytoplankton trap the energy of the Sun (primary producers)

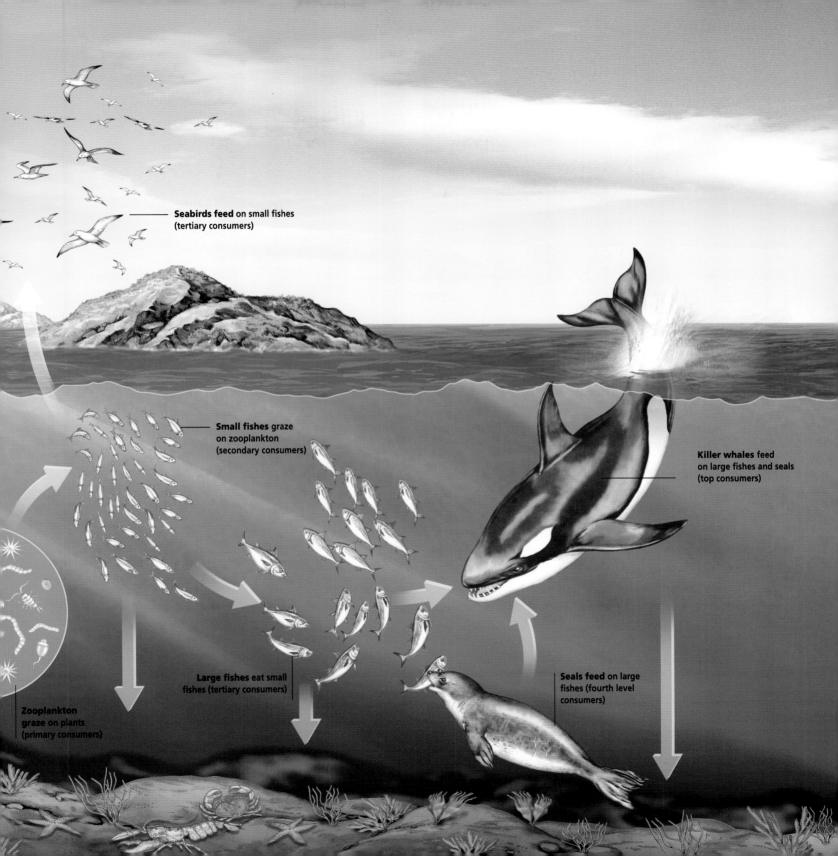

Seabirds feed on small fishes (tertiary consumers)

Small fishes graze on zooplankton (secondary consumers)

Killer whales feed on large fishes and seals (top consumers)

Large fishes eat small fishes (tertiary consumers)

Seals feed on large fishes (fourth level consumers)

Zooplankton graze on plants (primary consumers)

Plankton

The term plankton denotes the microscopic plants and small animals that spend all or part of their lives floating in midwater, drifting passively in the currents and eddies of the world's oceans. Planktonic plants (phytoplankton) are mainly single-celled algae. Because they require sunlight to live, they are restricted to the surface layers of the oceans, down to approximately 330 feet (100 m). Planktonic animals (zooplankton) are found at all depths, grazing on phytoplankton or plant and animal remains that sink down from the surface. Zooplankton consist of two types of animal: those that live out their lives in the water column (holoplankton); and those that spend part of their early lives feeding and drifting in the water before developing into the adult form away from their parents (meroplankton). Although minuscule, plankton is an essential component of the marine food web. Even top predators such as whales are dependent on the photosynthetic activities of phytoplankton to capture the energy of the Sun.

→ **This delicate grazer** of phytoplankton, with long, thin legs to slow its sinking, is the larva of a carnivorous lobster. These larvae often grow to 4 inches (10 cm) across the span of their legs before they settle and change to their adult forms.

PLANKTON DISTRIBUTION
This false-color satellite image of the world's oceans shows the distribution of phytoplankton in the surface waters between April and June. The colors represent different densities, from red (most dense), through yellow, green and blue to violet (least dense). Gray areas represent gaps in the data. The distribution image also highlights the relatively nutrient-poor areas in mid-ocean. In the northern hemisphere, the phytoplankton are at their most dense as the annual "spring bloom" spreads across the northern Atlantic and Pacific. The bloom is the result of nutrient buildup in winter and is triggered by the longer daylight hours of spring and increased light intensity. It subsequently fuels zooplankton blooms.

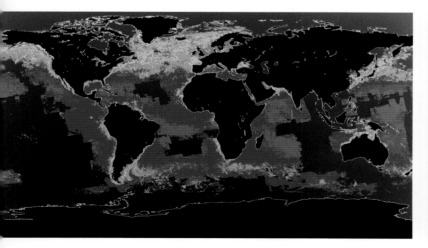

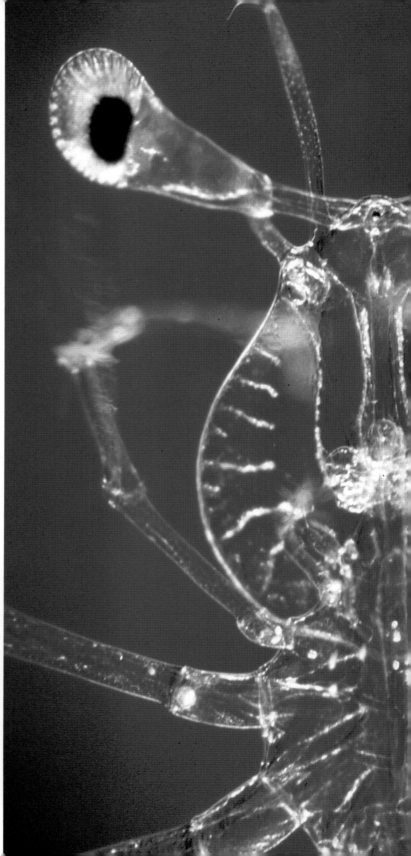

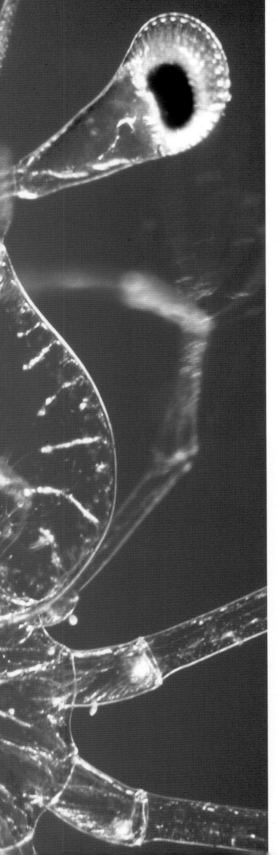

→ **This water flea,** or cladoceran, lives out its life in the water column, and even carries its offspring within it until they are able to fend for themselves. Its body is buoyant, so it uses energy only to move up and down.

↓ **Marine species whose adults** can move only limited distances make use of planktonic larvae to disperse offspring into new areas. This veliger is the larva of a marine snail that uses long extensions of its foot, covered in millions of beating microscopic hairs (cilia), to hang gently in the water.

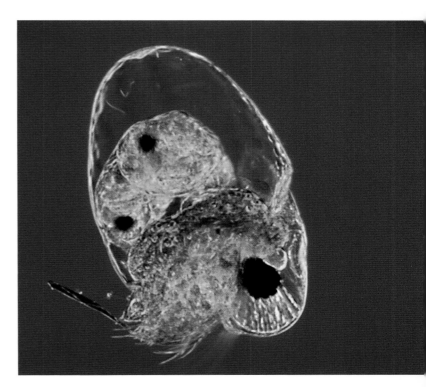

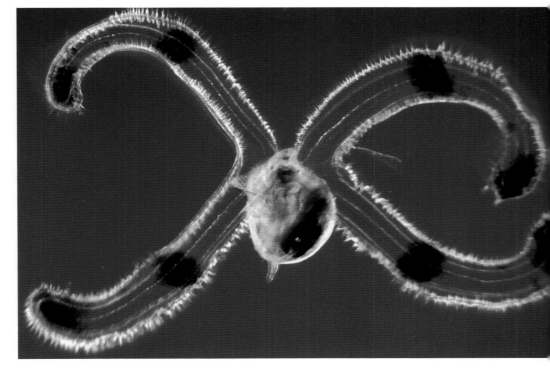

Seaweeds and flowering plants

Seaweeds are marine algae; they differ from land plants in lacking roots, flowers, seeds and fruit. They anchor themselves in place by means of a suckerlike structure called a holdfast. The simplest forms of seaweed are the green algae that consist of strings or sheets of algal cells; as their name suggests, they are often a vivid green. Brown algae include some of the largest, most complex forms of seaweed, such as the giant kelp that grows to great lengths and forms dense forests offshore. Their pigments are more sensitive to low light levels and enable them to live in deeper water than most other seaweeds. Red algae are the most diverse of the seaweeds in terms of number of species. They range from large fronds to encrusting, coral-like forms and include edible species, such as those used in Japanese cuisine.

The flowering plants are land plants that have adapted to life at the edge of the sea, or even in it. They range from those that are just salt tolerant and live above the splash zone, to sea-grass beds and mangroves that form some of the richest habitats in shallow coastal waters.

PHOTOSYNTHESIS

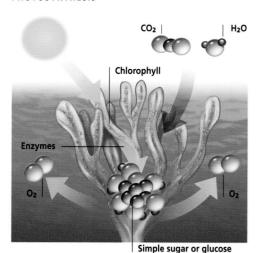

CO_2 H_2O

Chlorophyll

Enzymes

O_2 O_2

Simple sugar or glucose

← **Photosynthesis is a process** by which green plants transform solar energy into chemical energy. Light energy is captured by specialized molecules in plants, the best known being the chlorophylls. It is transferred to biochemical pathways and enzymes that convert water and carbon dioxide molecules into sugar molecules, at the same time releasing oxygen.

↓ **Sargassum, a brown seaweed**, is covered with numerous gas-filled, grape-shaped floats that enable the plant to live on the surface of the ocean. This seaweed provides a floating refuge for many fishes and invertebrates, such as these sargassum fishes. The greatest concentration of the weed is in the Sargasso Sea, where current systems create a gyre where the weed accumulates.

← **The giant kelp** is a brown seaweed that can grow 20 inches (50 cm) per day and reach 100 feet (30 m) in length. The plant consists of a holdfast and long, flexible, but tough, stems held up by gas-filled bladders at the base of the blades (leaves). To aid the transport of sugars produced in the blades by photosynthesis, giant kelp have developed specialized "trumpet cells" that are similar to the system in more advanced plants.

↗ **Mangroves are trees** or shrubs living on the fringes of tropical and subtropical seas. They thrive in saltwater because their leaves are adapted to excrete excess salt and prevent water loss. Mangrove roots are adapted to life in soft, oxygen-poor (or anoxic) mud by having prop, or buttress, roots that anchor the plants and spread out their weight. Above the mud, hollow tubes known as pneumatophores carry air down to the roots.

→ **Green seaweed is found** in the upper parts of sheltered rocky shores in temperate areas. Its pigments are more stable in strong sunlight than those in species farther down the shore. Green seaweed is able to cope with water loss and shrinkage that would irreparably damage the cells of less-tolerant species. Also, it can tolerate freshwater from streams.

Sponges

Sponges are among the simplest forms of sea creatures. They are little more than a loose association of cells organized into filter-feeding structures, and are found attached to the seafloor or to other marine surfaces. The main groups of sponges are classified according to the amount of folding of the body, which is one of the few consistent features in sponge body forms. Sponge cells exhibit some remarkable properties. If a sponge is completely broken up and the resulting slurry is left for a few hours, the cells will reorganize into a similar, although not identical, body form. Sponge cells are bound together by fibers of an elastic protein known as spongin. The body is given greater rigidity by the production of embedded mineral spicules, made of calcium carbonate or silica, that enable some species to grow very large. Sponges vary from small, irregular forms to very large barrels and tubes.

INSIDE A SPONGE

All sponges have the same basic internal structure. Water is drawn into the sponge through pores (ostia) connected to a network of internal canals. The current is created by the beating of microscopic hairs on cells (or choanocytes) that are grouped together in feeding chambers. Food particles are trapped in these chambers and taken up by surrounding cells. Waste products are expelled with the filtered water that leaves the sponge via pores called oscula. A sponge is able to filter its own volume rapidly, usually every 4 to 20 seconds.

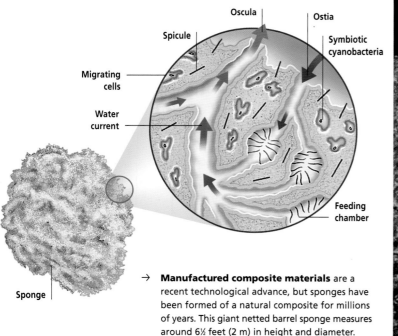

Oscula

Ostia

Spicule

Symbiotic cyanobacteria

Migrating cells

Water current

Feeding chamber

Sponge

→ **Manufactured composite materials** are a recent technological advance, but sponges have been formed of a natural composite for millions of years. This giant netted barrel sponge measures around 6½ feet (2 m) in height and diameter.

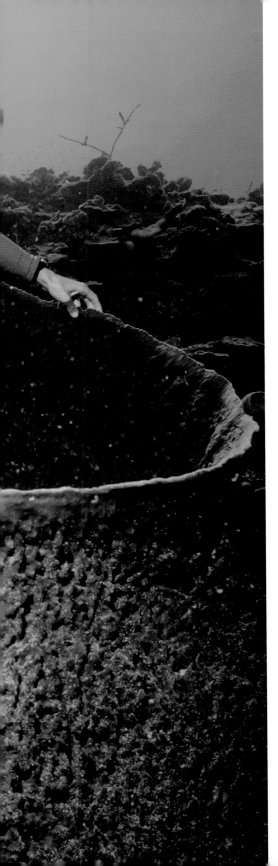

→ **These small sponges** have one of the simplest body forms. They have a purselike structure covered in small ostia. The ostia take water into the body of the sponge. The water is expelled via the large osculum at the top.

→ **This brilliant red** sponge has a more complex shape than many other species. It towers above the seabed and branches out to improve its chances of gathering food. Sponges are brightly colored as a warning to potential predators of noxious or toxic compounds that they contain. Many sponges found on reefs tend to resemble the surrounding corals in color and structure.

→ **Tube sponges** are one of the most common sponge varieties. They are found on reefs in a variety of vibrant colors. These sponges have evolved their distinctive shape to improve their feeding efficiency. Water flows over the top of the chimneylike opening, which helps to draw water in through the sides and reduces the amount of energy a sponge needs to feed.

Corals, jellyfishes and anemones

It is hard to believe a transparent jellyfish that is 98 percent water is closely related to animals who build massive reefs so hard they can tear the bottom out of a ship. Jellyfishes, corals and anemones are all known as cnidarians. They exhibit two body forms: the polyp, characteristic of corals and sea anemones, and the medusa form of jellyfishes. Cnidarians have stinging cells that capture prey or pick particles from the water. Most species are harmless to humans, but a few have venom that can cause severe injury or death. The reef-building corals live in colonies and secrete a hard calcium-carbonate skeleton around themselves. Over centuries, successive generations have gradually formed massive reefs. Coral reefs need clear water where sunlight can penetrate to the bottom to allow the process of photosynthesis to occur. Polyp colonies that lift themselves up off the sea bottom on a branching protein skeleton are known as sea fans. The highly prized precious or red coral is a sea fan that has pink or red chalky material embedded in the protein skeleton.

↓ **The distinctive bright** colors of corals are produced by algal cells (zooanthellae) living within the polyps. These algae use sunlight and nutrients from the coral to photosynthesize sugars, some of which pass back to their host. Corals also feed by capturing food and absorbing nutrients directly from the water.

Sea anemones are large muscular polyps. Their often colorful body consists of a hollow trunk that is attached, but not permanently fixed, to a hard surface by a disk-like foot. Food is caught by a crown of tentacles and digested in the cavity within the trunk. Any undigested material is ejected through the mouth.

INSIDE A STING CELL

Sting cells can be triggered either by direct contact or by chemicals released by the prey or potential predator when in close proximity to the sting cell. Once triggered, the fluid in the cell rapidly absorbs water, and muscle fibers around the cell contract. This combination of events causes a sudden increase in the internal pressure of the cell, and forces fluid up the hollow center of the venom-tipped thread coiled within. The inflating sting bursts out of the top of the cell. It does so with sufficient force to drive its fine tip into the prey's skin and inject venom. Most jellyfish species are harmless to humans, but the toxicity of a few can cause death.

Jellyfish medusae are often found in huge swarms in both tropical and temperate waters. The ringlike structures that hang down inside the bell are mouths. They are fed by the fringes of tentacles. By eating fish larvae or the food larvae eat, large jellyfish swarms can seriously affect zooplankton numbers and damage fish stocks.

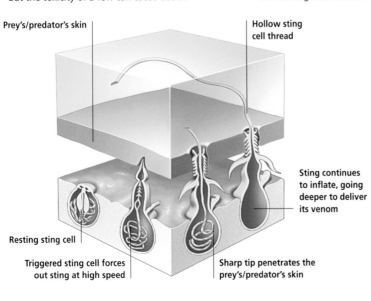

Prey's/predator's skin

Hollow sting cell thread

Sting continues to inflate, going deeper to deliver its venom

Resting sting cell

Triggered sting cell forces out sting at high speed

Sharp tip penetrates the prey's/predator's skin

Comb jellies

Comb jellies, or ctenophores, are radially symmetrical and look like true jellyfishes. However, they are a completely separate group, consisting of approximately 100 species, and are exclusively marine. Comb jellies have a transparent, gelatinous body that, under normal lighting, is difficult to see. Often, the only visual indications of their presence are eight rows of cilia over the body that beat in coordinated waves. These are referred to as ciliary combs and they refract light, giving rise to a shimmering rainbow of colors as they beat. Comb jellies range from minuscule to up to 7 feet (2 m) long. All are carnivorous and most feed on zooplankton caught using two long tentacles. The tentacles are covered in sticky, or lasso, cells, which hold on to prey. Only one rare species is known to have sting cells. Most species are active swimmers and live in the water column. A small number, the purse jellies, live on the seabed.

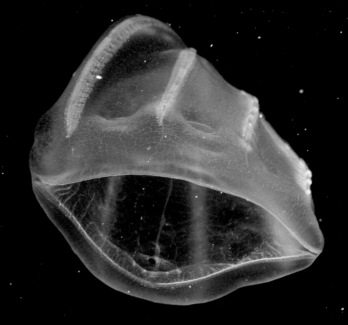

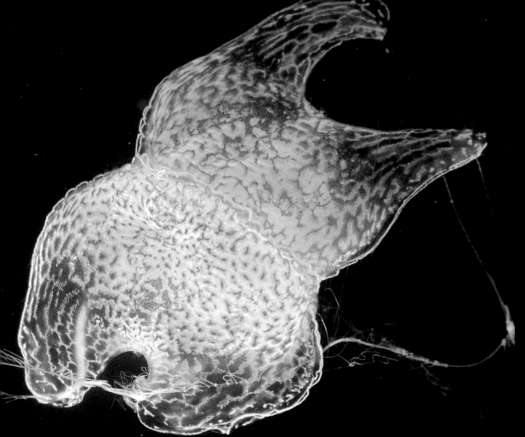

↑ **This species belongs to the group** of comb jellies that have completely lost their long feeding tentacles. They are voracious predators that feed almost exclusively on other species of comb jellies.

← **Two purse jellies mate on the seafloor.** They are hermaphrodites, and produce both eggs and sperm. These eggs and sperm come together in the water column, so fertilization is external. The resulting larvae drift off in the current to become temporary plankton.

↓ **This comb jelly, or sea gooseberry,** has a relatively small body that is approximately ½ inch (1.5 cm) long. Its two long feeding tentacles, which can be many times its body length, have cells called colloblasts that fire sticky threads and aid in catching prey.

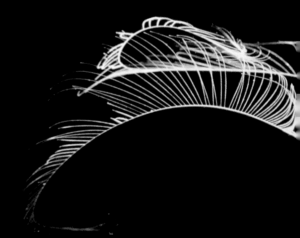

← **Venus's girdle is a long** ribbon-shaped comb jelly that is often more than 4 feet (1.5 m) long and about 3 inches (7.5 cm) wide. Its body is transparent and is usually tinged a delicate purple. Although it has a well-developed musculature and is able to swim with a snakelike undulating motion, it usually hangs motionless and vertical in the water column, waiting to ambush prey. Venus's girdle captures plankton using small tentacles along its length; each group of tentacles directs food toward the mouth. In daylight hours, this animal is detectable only by the golden shimmer emanating from its beating tentacles. At night, as with a number of other comb jellies, it is phosphorescent and emits light if disturbed.

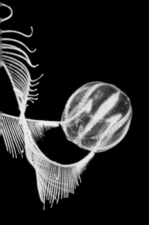

INSIDE A COMB JELLY

The illustration of a comb jelly (*right*) shows the characteristic bands or combs of cilia and the two long feeding tentacles. The gelatinous body of the animal is neutrally buoyant—it does not fall or float up. This is because approximately 96 percent of its body is water. By controlling the rate of beating in the combs that are equally spaced around its body, the comb jelly can either hang motionless waiting to ambush prey or move very precisely in three dimensions. The mouth is at the base of the body and the anus on top. The lines of the vascular canals are also visible in this illustration. These canals distribute nutrients absorbed from food in the gut to the rest of the body. Research suggests that they also improve the distribution of oxygen to the cells and carry waste products back to the gut for removal.

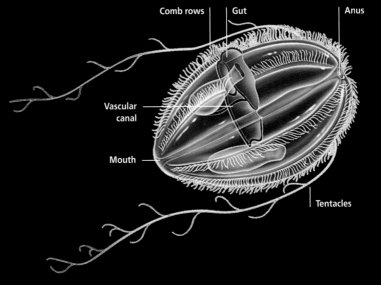

Comb rows | Gut | Anus

Vascular canal

Mouth

Tentacles

Marine worms

The simplest marine worms visible to the naked eye are the flatworms (platyhelminths). They have flat bodies symmetrical along their long axis, and a head with well-developed sense organs. However, they have a blind-ended gut like jellyfishes and sea anemones, and they do not have gills or a blood system. Many tropical species are brightly colored to warn potential predators that they contain noxious chemicals. Some can even extract undischarged sting cells from sea anemones for use in their own bodies. The ribbon, or nemertean, worms are similar to the flatworms but have a gut with a mouth and anus, enabling them to feed more efficiently and to grow many times larger. They are active carnivores, hunting small animals buried in soft mud and sand. One of the most remarkable characteristics of the ribbon worms is their ability to change length. The bootlace worm found on northern temperate shores is often 33 feet (10 m) long when moderately contracted, but it can expand to more than twice that length. Most marine segmented, or annelid, worms are in a group known as the polychaetes. These have a very wide variety of body forms and lifestyles. Some are active predators above or below the sediment surface; others are fixed in tubes and filter feed. All have bodies composed of repeating segments. The peanut, or sipunculan, worms are probably distant relatives of the annelids, but they lack segmentation and have well-developed organ systems.

INSIDE A TROCHOPHORE

The trochophore larva (*right*) is a form used by a number of worm groups. The larva moves through the water column using an equatorial ring of cilia (prototroch), feeding on phytoplankton. Other bands of cilia, such as those around the anus and below the prototroch, provide some directional movement and also help to reduce the trochophore's sinking rate. This larva uses the apical tuft of cilia on top to orientate itself and detect water movements. It will eventually elongate until a miniature adult is formed and the trochophore is absorbed into the head.

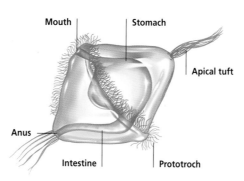

Mouth | Stomach | Apical tuft | Anus | Intestine | Prototroch

← **These Caribbean fanworms** have a delicate crown of feathery white tentacles covered with sticky mucus that is used to trap particles in the drifting water. The material is sorted and transferred down to the mouth by specialized groups of cilia. When danger threatens, the tentacles can be rapidly withdrawn into the tube containing the rest of the body.

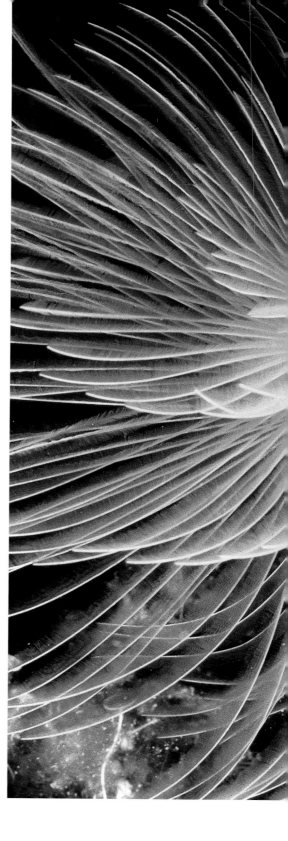

← The spiral tentacles on this polychaete tubeworm are used for feeding and respiration. The worm is raised up by the chalky tube it has secreted around itself and which it continually adds to as it grows.

← The zebra flatworm has a warning coloration that is found in many animals. Bold cream stripes on black are a recognized signal to a predator that the bearer may be unpleasant, if not poisonous.

↓ The bearded fireworm derives its name from the distinctive tufts of fine spines on each side of every segment of its body and from its ability to eat firecoral polyps with no ill effects.

Lophophorates

Sea mats, lamp shells and horseshoe worms are three phyla of mainly marine invertebrates that are known collectively as lophophorates. They all have an unusual filter-feeding structure, called a lophophore, that looks like a curved or coiled brush. These animals have different body forms and life histories, but they also share many characteristics.

Sea mats, or bryozoans, are the largest group of lophophorates. There are some 5000 species. Sea mat colonies are comprised of individuals called zooids that collaborate to make a skeleton in which they live. This can be either an encrusting patch on rocks or seaweeds, or lacy colonies that resemble bleached seaweed. The lamp shells, or brachiopods, resemble ancient pottery lamps or bivalves, but the presence of a long fleshy stalk and the lophophore indicates their true identity. The horseshoe worms, or phoronids, get their common name from the distinctive horseshoe shape of their lophophore. They live worldwide in tubes built of sand grains that are attached to hard surfaces in shallow water.

← **Horseshoe worms** use the whorls of their delicate tentacles to trap particles from the drifting water. This material is then pulled down to the base of the tentacles and is sorted into edible and non-edible items by the lophophore.

→ **A sea mat grows** with anemones on a kelp frond. Although each individual zooid has a separate compartment, it is connected to its immediate neighbors by strands of soft tissue.

← **Tufts of tentacles** emerge from the tops of compartments, each containing one sea mat zooid. The tentacles form the lophophore in sea mats and have three bands of cilia down the sides and middle of each tentacle. These bands "bounce" particles so they move to the mouth at the base of the tentacles.

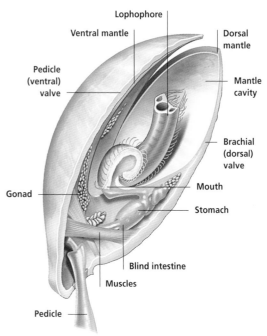

Lophophore

Ventral mantle

Dorsal mantle

Pedicle (ventral) valve

Mantle cavity

Brachial (dorsal) valve

Gonad

Mouth

Stomach

Blind intestine

Muscles

Pedicle

INSIDE A LAMP SHELL

Most lamp shells have a fleshy stalk, or pedicle, that either secretes a cement that sticks them to rock, or pushes into mud to act as an anchor. In order to feed, complex sets of muscles open the shell so the lophophore can create a water current to bring in food and oxygen and wash away waste products. Many lamp shells take up oxygen by having a pigment called hemerythrin in their blood.

Lamp shells only reproduce sexually. The eggs are fertilized in the water column and develop into larvae.

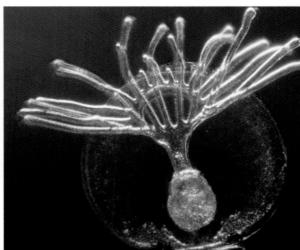

Mollusks

Mollusks are a diverse group of animals whose origins lie in the ancient seas of the Precambrian era more than a billion years ago. Today, the most primitive form of marine mollusk is the chiton. Its body is covered by eight interlocking, hinged chalky plates. Gastropods—sea slugs, snails and limpets—are the largest group of mollusks. They have, or have had in their evolutionary history, a single calcareous shell. The sea slugs lost their armorlike shell over time and rely on warning coloration and chemical defenses. Sea slugs are often colorful and have feathery gills or other ornamentation on their backs. Snails and limpets have shells in a great variety of sizes, shapes, colors and degrees of ornamentation. A small number of gastropods adapted to life in the water column. They either extended their foot into a winglike structure, as in the sea butterflies, or, like the purple sea snail, secreted a raft of bubbles from which to hang and catch their jellyfish prey. Bivalves are a group of mollusks that have evolved a hinged shell and have largely abandoned locomotion. Most bivalves are filter feeders. A few, such as the ship worm, are borers that feed on wood and detritus. The most advanced mollusks are the cephalopods, which include octopus and squid. These active hunters use their acute senses and camouflage to ambush prey.

→ **The giant clam is found** in shallow tropical waters. The spotted green coloration is caused by green symbiotic algae. The light spots are lenslike structures, derived from simple eyes, that the clam uses to focus sunlight on the algae.

↓ **This colorful tropical scallop** has a fringe of sensitive tentacles to detect anything attempting to attack its exposed soft tissues while it is filter feeding. Any stimulation of the tentacles will cause the scallop to snap shut.

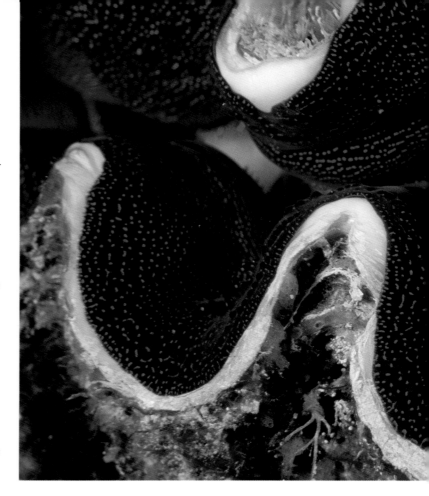

This sea slug is able to maneuver its soft, highly flexible body to search for food around the spines of a sea urchin. The urchin's spines and the warning coloration on the slug are effective deterrents to potential predators.

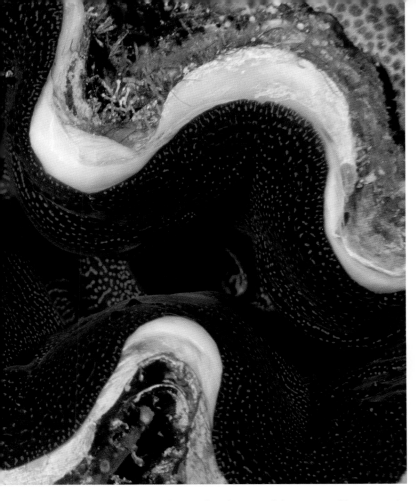

CEPHALOPOD SUCKERS

Many cephalopods capture prey by shooting out a pair of tentacles bearing spiked suckers, often with serrated tips. These structures ensure that even the most slippery prey is unlikely to escape being drawn toward the venomous beak that administers the final *coup de grâce*. The size, shape and imprint of the suckers are often characteristic of particular species. The existence of many deepsea squid species was first deduced from sucker marks on deep-diving whales, long before any squid specimens were found. For example, the colossal squid (*Mesonychoteuthis hamiltoni*) was first named in 1925, partly by identifying its sucker scars on whales. A complete specimen of this squid was not found until 2003. This illustration shows the club at the end of the squid's tentacle. Squid also have eight arms, which are much shorter than tentacles, but also carry suckers.

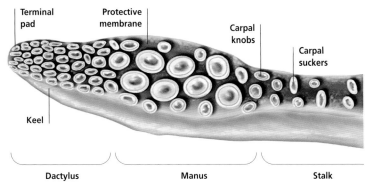

The blue-ringed octopus is notorious because of the potency of its neurotoxic venom. It can kill an adult human in minutes. These small cephalopods rely on the toxin to quickly subdue prey that could easily damage their soft bodies.

Cephalopods use their nervous system to flash changing colors and patterns over their bodies. It is thought that these patterns are a form of communication that is necessary for social interactions, such as the cuttlefish courtship shown here.

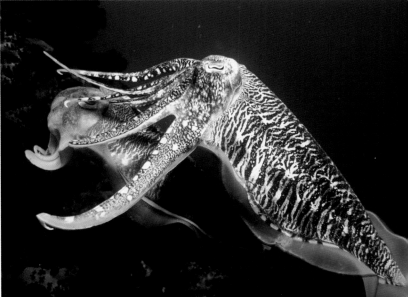

Crustaceans

Arthropods are the largest phylum in the animal kingdom. In the sea, they are mainly represented by the crustaceans. The common features of all the arthropods are a hard, often armored, external skeleton, or exoskeleton, jointed legs and feeding appendages. The exoskeleton provides the animal with a high degree of protection, and has allowed the development of powerful muscles. It limits the creature's size, however, because it has to be lost periodically by molting to allow the soft tissue to expand and grow.

In the oceans, crustaceans have gills for respiration and an exoskeleton reinforced with calcium carbonate absorbed from the surrounding water. There may be as many as 45,000 species of crustaceans and they are found in all major marine habitats. Scientists regard them as the marine equivalent of the dominant terrestrial arthropods—the insects.

→ **Young spiny lobsters** undertake an annual migration from juvenile to adult feeding grounds.

As larvae, gooseneck barnacles are similar to other crustaceans, but are dramatically different as adults. Most of the animal is encased in tough, chalky plates, with feathery legs that strain food particles.

The spider crab has long spines to deter predators. Its legs have sharp tips, enabling it to climb soft coral. Pincers on the forelimbs help it pick up particles of food, including pieces of soft coral.

Despite its name, this pink squat lobster is more closely related to hermit crabs. It uses its sharp claws to pick up morsels of food from the seabed and to defend itself, usually in territorial disputes.

↓ **By spreading its weight** between its walking legs, this crab is able to move over soft sand. It scoops up sand and mud with its first pair of clawed limbs and uses its mouth parts to sift out any organic material.

↓ **This tropical shrimp** is picking its way over the tentacles of a sea anemone. Its delicate limbs must avoid triggering the anemone's lethal sting cells. The shrimp's distinctive color is produced by pigments in the exoskeleton.

This shrimp has prominent compound eyes. That is, its eyes are made up of individual units that give a mosaic of images over a wide field of vision and are good at detecting movement of prey or predators.

Echinoderms

The echinoderms—starfishes, feather stars, sea cucumbers and sea urchins—form one of the largest groups of invertebrate animals in the sea. There are approximately 60,000 species, all of which live in marine habitats—from warm tropical waters to icy polar seas, and from the shoreline to the bottom of the deepest ocean trenches. The body forms of all echinoderms are variations on a simple theme. They are radially symmetrical and have five arms, although these may be lost or modified while they develop. The basic body plan is seen in the starfish. The feather stars can be thought of as starfishes living upside down, with the upturned arms reduced to flexible, feathery stalks that beat in the water. The sea cucumbers have fused their arms to form a cylinder. Their soft, leathery bodies enable them to move in small crevices. Sea urchins have fused their arms to form a hollow disk of hard, chalky plates. Most echinoderms feed on small particles in the water, but some are active predators of bivalve mollusks and corals.

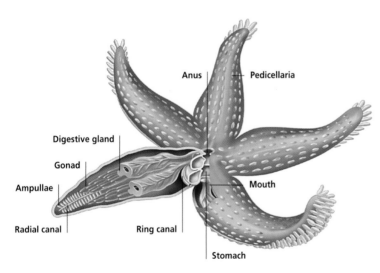

Anus | Pedicellaria

Digestive gland

Gonad

Ampullae

Mouth

Radial canal | Ring canal

Stomach

INSIDE A STARFISH

Echinoderms have an internal skeleton covered by soft tissue, often with spines or lumps. In many species, there are also pincer structures called pedicellariae that remove debris. The central disk has a mouth on the underside and an anus on the upper surface. A water vascular system is used to operate the tube feet by hydraulic pressure. The tube feet have sucker tips and can hold prey or grip surfaces when the animal moves. They are strong enough to pull apart a live mussel.

→ **Starfishes are often found** in large aggregations. They come together either to feed on an abundant food source, such as a mussel bed, or to bring males and females together for spawning.

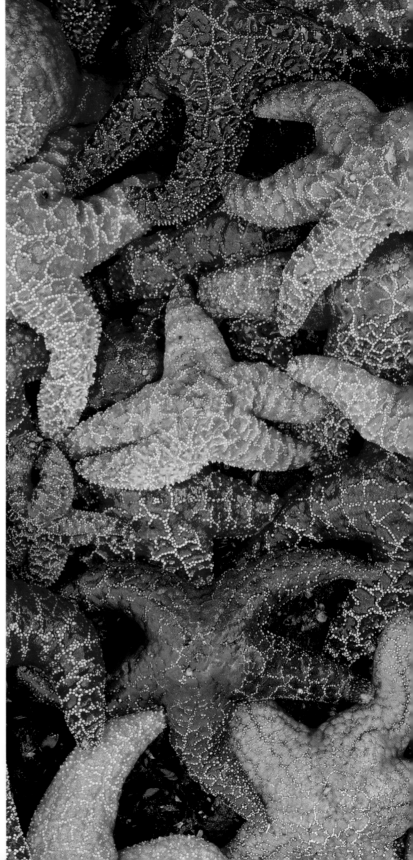

→ **Feather stars are exclusively** suspension feeders. The delicate arms have evolved to catch small plankton and food particles suspended in the water. A complex arrangement of cilia and grooves sorts the particles. The edible material is passed down to the mouth at the base of the arms.

↓ **The thick spines** on the slate pencil urchins will deter all but the most intrepid predator. The spines also provide shelter for small fishes.

↘ **Sea cucumbers** have no spines. They rely on other forms of defense. Their bright colors act as a warning that they contain toxic chemicals. Many species fire their internal organs out of their mouth or anus to distract a predator.

Sea squirts and lancelets

Surprising as it may seem, sea squirts and lancelets are our distant cousins, as they are in the same phylum (chordates) as humans. All 49,000 species of living chordates share three important characteristics. During development, each one has a single, hollow nerve chord, gill slits and a flexible rod (notochord) to support the nerve cord. What distinguishes humans and other animals that are described as vertebrates is that the notochord is replaced by, or incorporated into, a true jointed backbone. Sea squirts and lancelets do not have a backbone at any stage so are technically invertebrates. Adult sea squirts are filter feeders. Water is drawn into the mouth, or inhalant siphon, by the beating of cilia that cover a sievelike basket at the front end of the gut, and the openings are derived from gill slits in the larva. Food particles are trapped on the mucus that covers the sieve and are eventually drawn in a string into the gut. Filtered water and waste pass out of the animal through a second exhalant siphon. Some sea squirts are found as colonies on hard surfaces. The salps are sea squirts that have gone back into the water column, either as individuals or as ribbonlike colonies. The lancelets are invertebrate chordates found in soft sediments in shallow water.

INSIDE A SEA SQUIRT TADPOLE

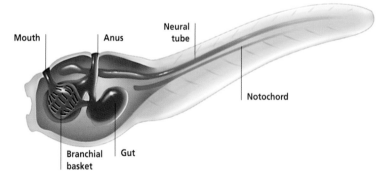

Mouth | Anus | Neural tube

Notochord

Branchial basket | Gut

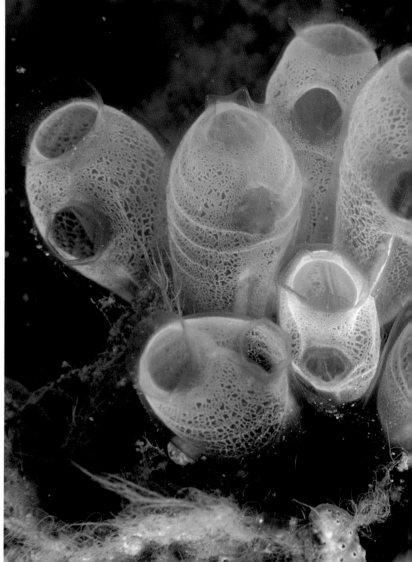

↑ **The sea squirt tadpole larva** has a well-defined head, often with an eye, and a muscular tail for swimming. The larval stage is short. The tadpole larva does not feed; its sole purpose is to find a suitable site to settle. Once this is found, the tadpole larva uses the adhesive papillae on the head to secure itself to a hard surface. The notochord and tail are rapidly absorbed, leaving the front end to expand into the simple, filter-feeding adult form.

→ **In this image of a lancelet tail,** the notochord and the nerve cord running down the mid-line, with V-shaped muscle blocks to either side, are clearly visible. The contraction of these muscles produces an S-shaped swimming motion.

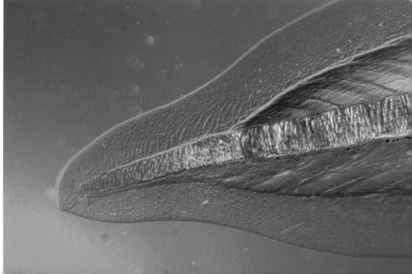

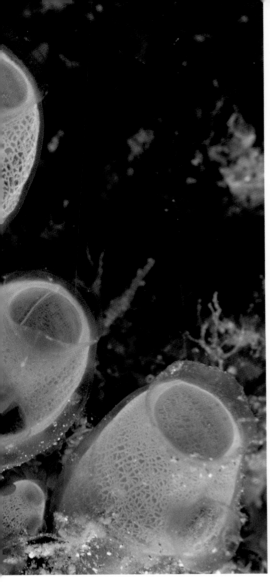

← **In these colonial sea squirts,** water is drawn into the colony through small holes on the side, and leaves through the large single hole on top. Their bright colors advertise the fact that they contain noxious chemicals. Potential predators also have to deal with the sulfuric acid and the metal vanadium found in the blood of all sea squirts.

←← **The large holes** on the tops of sea squirts are their inhalant and exhalant siphons. There are fringes of spines and sensitive tentacles around the hole to prevent unwanted materials entering. Any disturbance causes the siphons to close. Sea squirts get their name from their ability to rapidly eject unwanted material from their body.

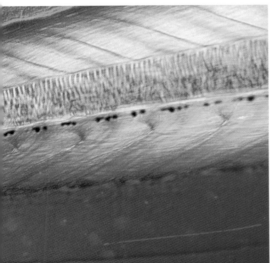

← **Each of the "flowers"** in this photograph is a group of individual sea squirts, sometimes known as polyps. The sea squirts are embedded in a semi-transparent rubbery matrix that covers hard surfaces. They are often found on sheltered rocky shores or reefs, where wave action stirs up food particles and ensures the water is well oxygenated.

Lampreys and hagfishes

Lampreys are found in temperate freshwaters of the northern and southern hemispheres, but some species spend part of their adult lives in the sea. Adult lampreys can be either non-parasitic or parasitic. The non-parasitic species are called brook or dwarf lampreys. After a larval stage lasting as long as seven years, they do not feed before spawning and dying. The parasitic species may live for a further three years after metamorphosis and grow to 3 feet (90 cm). These parasitic forms stay in freshwater or migrate to the ocean.

Hagfishes are exclusively marine in temperate zones but also occur in the cool, deep waters of the tropics. Unlike lampreys, hagfishes do not have a larval phase but produce large, yolky eggs that hatch into miniature versions of the adult. They are nocturnal predators of small invertebrates and scavenger feeders. Feeding is helped by their ability to tie themselves in a knot, and they use this as a form of leverage to tear off flesh. The jawless but biting mouth is surrounded by six barbels. A terminal nostril detects dead or dying prey.

→ **Lampreys spawn in freshwater** after constructing a nest pit by moving stones with their sucker mouths. The female attaches herself to a rock, the male wraps his body around her, and sperm and eggs are released simultaneously.

↓ **Hagfishes have degenerate eyes** which are replaced by two white patches of light-sensitive cells on their head. They are particularly well known for their ability to produce vast amounts of mucus (literally bucketloads) in just a few seconds—hence, their alternative names of slime eels, slime hags or snot eels.

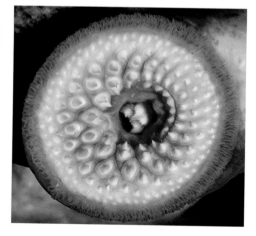

The sea lamprey has a circular mouth and an oral disk with horny teeth. It uses its sucker mouth to anchor itself to its host, consuming body fluids, blood and some flesh. The arrangement of the teeth is an aid to identification.

Hagfishes are mainly nocturnal predators of small invertebrates, but they are also scavengers. They are able to produce vast amounts of mucus that help them penetrate animal carcasses, such as dead fishes and the remains of whales and dolphins. They also have long, thin, extremely flexible bodies.

ANCIENT SURVIVORS

Although direct linkages are impossible to make, many scientists have suggested that lampreys and hagfishes are the sole surviving representatives of the earliest jawless vertebrates that appeared over 500 million years ago. Unlike all other living vertebrates, lampreys and hagfishes do not possess jaws and so are placed in a group called the Agnatha (meaning "without jaws"). The very first agnathans in the fossil record had a bony external skeleton; the lampreys and hagfishes are quite different and do not have any bone. They are eel-like, with porelike gill openings, and lack both scales and paired fins. There are thought to be 66 species of hagfishes and 39 species of lampreys.

→ **This sea lamprey has attached itself** to a fish with its sucker and consumed so much of its blood and body fluids that the fish has died. The lamprey can extract blood up to 30 percent of its own weight each day. Large hosts can survive a lamprey attack but smaller prey, such as herring, frequently die.

Sharks, rays and chimaeras

Sharks, rays and chimaeras are cartilaginous fishes (Chondrichthyes). There are thought to be 415 species of sharks, 547 species of rays and 37 species of chimaeras. The largest fishes in the world are the plankton-eating whale shark (greater than 40 feet/12 m) and basking shark (30 feet/9 m). There are also large carnivorous sharks, well known to humans, such as the white, tiger, hammerhead and thresher (15–20 feet/5–6 m). On the other end of the scale, deep-water dogfishes grow to less than 8 inches (20 cm). Rays can also achieve great size— manta rays can grow to a width of 23 feet (7 m). Most sharks and rays are entirely marine but 28 species are found in freshwater. Although often portrayed as primitive, modern sharks and rays are highly specialized fishes and quite different from their early ancestors. The chimaeras are entirely marine and reach a maximum length of about 5 feet (1.5 m).

REPRODUCTION

All living sharks, rays and chimaeras have internal fertilization and produce large, yolky eggs in relatively small numbers. This makes them highly susceptible to over-exploitation. Males do not have a true penis, but have modified pelvic fins known as claspers to help the transfer of sperm; chimaeras also have an additional clasper on their heads. Females either lay eggs in tough leathery pouches or nourish embryos internally for several months before giving birth. Gestation can be extremely long. The spiny dogfish has a gestation period of two years, while the basking shark's may be more than three years. These fishes have a range of reproductive strategies, varying from simple egg laying (oviparity) to advanced live bearing (viviparity), where the young are nourished via a placenta similar to that of mammals.

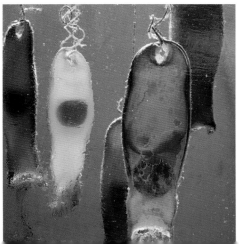

Gray reef sharks are an active, social species that feed on small fishes, squids, octopuses, lobsters and crabs. Along with whitetip and blacktip reef sharks, the gray reef is one of the three most common species of reef shark in the Indo-Pacific. Although only a moderately sized shark growing to 8 feet (2.4 m), this species will attack divers if threatened. It has a distinct threat display that consists of an exaggerated swimming pattern. The shark then either flees or attacks and bites with tremendous speed.

←← **The southern stingray** is often found lying partially buried in sand beneath warm water. As in other stingrays, venom is delivered through spines at the base of the tail.

← **Shark and ray egg cases** are comparatively large and are protected by a tough outer covering. These swell shark egg cases protect the embryos for 7½ to 10 months, from laying to hatching. The young emerge when they are about 6 inches (15 cm) long. Each egg case contains a single embryo, and is attached to seaweed or the seafloor by strong, coiled tendrils.

→ **Chimaera, or rabbitfish,** such as this one, spend their entire lives close to the seafloor, where they feed on invertebrates. They are often found in deep water, and can reach depths of up to 3300 feet (1000 m). In northern areas, they make an annual summer migration, moving inshore to depths around 130 to 330 feet (40 to 100 m). Slow-moving fishes, they are most often seen in small groups.

Bony fishes

Bony fishes are the most abundant and species-rich group of all the vertebrates. Unlike the sharks and rays, they have skeletons made of bone. Most living species of bony fishes belong to the ray-finned fishes (Actinopterygii) because the fins have fin rays that are soft and often branched. The ray-finned fishes can be further divided, depending on the presence of spines, into the soft-rayed and spiny-rayed fishes. There are four main groups of living soft-rayed fishes, three of which are marine. These are the bonytongues and relatives (freshwater), the eels and tarpons, the herrings and a number of sub-groups that include catfishes. There are at least 24,600 species of ray-finned fishes, of which 11,345 have soft fin rays and the remainder have spines as well as soft fin rays. The internal anatomy of the soft-rayed fishes differs in two main respects from that of spiny-rayed fishes. Firstly, where it is present, the gas bladder in soft-rayed fishes remains connected to the esophagus and gas can be lost or gained by swallowing air or releasing gases via the mouth. There is also gas secretion and reabsorption via the gas gland that is attached to the gas bladder. Soft-rayed fishes also differ from spiny-rayed species in having a separate pancreas and liver.

↘ **Northern anchovies inhabit** coastal waters, mainly within 19 miles (30 km) of shore, but as far out as 300 miles (480 km). They form large, tightly packed schools that enter bays and inlets to feed on euphausids, copepods and decapod larvae, both by random filter feeding and by pecking at prey.

↓ **The blackspotted or laced moray eel** is found on reef flats and outer reef slopes of continental reefs in the western Pacific to East Africa. It lives in holes with cleaner fishes or shrimp, feeding on cephalopods and small fishes.

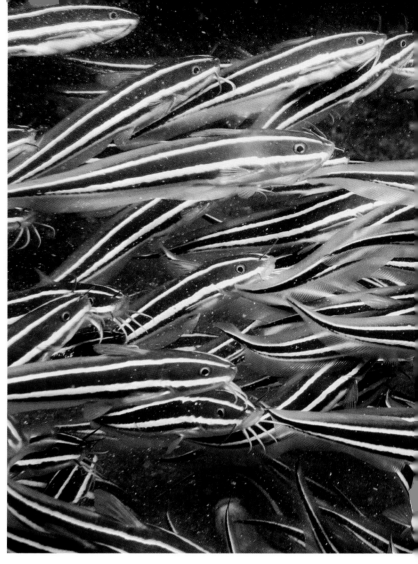

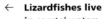 **Striped eeltail catfishes** occur around reefs in the Indian and western Pacific oceans. The juveniles live in tightly packed groups that swarm to deter and confuse predators. They have serrated and venomous spinelike rays on dorsal and pectoral fins which can inflict painful wounds. They are also called bumblebee catfishes because of the buzzing sound they emit when removed from water.

→ **The ribbon moray** is often partially buried in sand throughout the Indo-Pacific region. It grows to a length of 51 inches (130 cm) and feeds on small fishes. It undergoes abrupt changes in coloration and sex when functional males change sex to become females.

← **Lizardfishes live in** coastal waters throughout the warmer parts of the Atlantic, Pacific and Indian oceans, living partially buried in soft sediments or around rock or coral reefs. The head looks reptilian and the large mouth contains slender, sharp teeth. The largest species grow to 2 feet (60 cm) in length and are voracious predators of small fishes.

Spiny-rayed fishes

Most of the fish species living today belong to the 13,262 species of spiny-rayed fishes, of which 10,177 are marine. The marine forms can be divided into three main types: the mullets, the sand smelts and their relatives, and the perches and perchlike fishes. As their name implies, the unique characteristic of this group of fishes is that, in addition to soft fin rays to support their fins, they also have sharp spines in their dorsal, pelvic and anal fins. The spiny-rayed fishes are found in all marine habitats and have evolved a huge variety of lifestyles, often modifying their major body structures, such as their complex jaws. Most spiny-rayed fishes have a gas bladder that is sealed, so buoyancy is regulated solely by secretion or absorption of gases via the gas gland. Some species, such as some tunas, move rapidly through large vertical distances and avoid the difficulty of regulating their gas bladders by dispensing with them entirely. Spiny-rayed fishes have a combined pancreas and liver called a hepatopancreas.

→ **Seahorses are unique** in that the males have a specialized brood pouch. The female lays her eggs at the entrance to the pouch, where they are fertilized by the male. He cares for them and gives birth.

↓ **Creole wrasses live** on seaward reef slopes and shallow patch reefs in the tropical western Atlantic. These wrasses are normally found in large midwater aggregations that resemble shoals of damselfishes.

↗ **Coral cods live in clear water** around the well-developed coral reefs of the Indo-Pacific region. They are ambush predators that wait for smaller reef fishes to pass before making their attack.

⇒ **Blue marlin hunt** the Atlantic, western Pacific and Indian oceans in search of smaller fishes. They often weigh over 1000 pounds (450 kg) and reach speeds of up to 46 miles per hour (75 km/h).

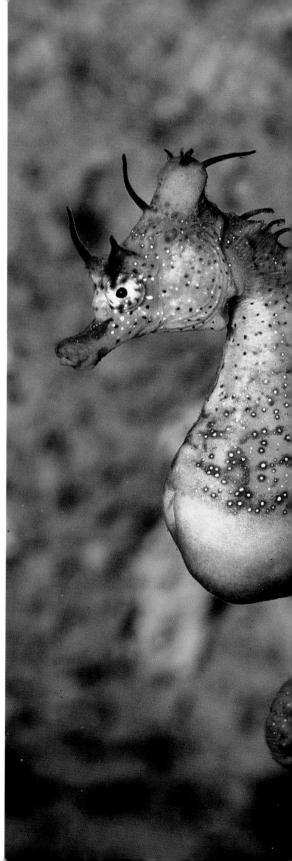

SEX ROLES

Hermaphrodites and sex reversals occur in several groups of fishes, most often in spiny-rayed species. Hermaphrodites that are simultaneously functional males and females are found in some species of sea bass. Sex reversals are common in species where there is strong territoriality and closed breeding groups known as harems. Wrasses, parrotfishes and damselfishes live in harems of small females that are controlled by a single, larger male which, if removed, is replaced by the largest female, who quickly becomes a normal male. Anemone fishes live in small groups of non-breeding individuals, controlled by a larger male–female pair. If the dominant female is removed, the dominant male takes her place and becomes a functional female.

Marine reptiles

Although reptiles are the most ancient, truly terrestrial vertebrates, they have also evolved marine species several times during their 300 million-year history. The turtles are the oldest group of living reptiles in the sea, though there are only eight truly marine species left. Sea turtles spend most of their lives in the water, coming ashore only to lay eggs. Sea snakes are relatively recent returnees to the sea but, nevertheless, they are the most abundant living marine reptiles, both in terms of numbers and variety of species. Although they are all highly venomous carnivores, related to cobras, sea snakes are not aggressive to humans. The size of their mouths, in which the fangs are placed to the rear, makes it difficult to be bitten. Lizards and crocodiles are represented by only two species, the marine iguanas and the saltwater crocodiles. The main natural limitation on the distribution of reptiles is water temperature; they are cold blooded and cannot survive for any length of time in cold water.

↗ **Adult female turtles haul ashore** on secluded sandy beaches once a year to lay eggs in the sand at the top of the shore. The hatchlings have no parental care and are preyed on as they make the perilous dash down the beach to the water.

→ **Green sea turtles are found** in warm coastal waters around the world, feeding on sea-grass beds. They reach a maximum length of 4 feet (1.2 m). Unlike other sea turtles, green sea turtles leave the water to bask in the sun, making them easy prey.

↓ **The banded sea snake swims** by using its oar-shaped tail. Unlike other sea snakes, it is amphibious. It hunts as if on land, crawling over the seabed around coral reefs and investigating crevices and holes for fishes and large crustaceans.

FARMING SEA TURTLES

Sea turtles have been hunted since antiquity for their shells and their meat, which is regarded as a delicacy. For centuries, live sea turtles were shipped to Europe for turtle soup. Turtle meat is still eaten in many parts of the world. To compensate for the decline in wild populations, various attempts have been made at farming sea turtles. The first farms raised them by taking eggs from breeding beaches, but this source became unreliable as the number of adults in the wild declined. Adults were then captured to act as broodstock, with varying degrees of success. In more recent times, farms have been set up that have tried to maintain self-sustaining groups of turtles, without recourse to either eggs or adults taken from the wild.

→ **The marine iguanas** of the Galapagos Islands feed on the algal fronds that grow subtidally around the islands' rocky shores. The waters around the Galapagos are cool. Iguanas bask on rocks in sunshine to ensure they have enough heat to swim underwater for just a few minutes at a time. They need to return to shore to warm up their bodies again.

Saltwater crocodiles pose a danger to humans and are more feared than sharks in some parts of the world. Attacks are relatively rare, however, with 14 deaths being recorded in Australia between 1976 and 2003. The crocodile's reputation for attacking humans and the demand for its skin increased hunting, causing a serious decline in the wild populations. However, since the 1970s, protective measures in Australia have saved these crocodiles from extinction.

Dugongs and manatees

Dugongs, or sea cows, are marine mammals that share an ancestor with elephants. They grow up to 10 feet (3 m) long and weigh up to 880 pounds (400 kg). The largest dugong, Steller's sea cow, weighed as much as 5 tons (5.5 t); it was hunted to extinction during the eighteenth century. The dugong has a rounded body with forelimbs modified into flippers and rear legs fused into a single tail fluke. Its thick skin, beneath which is a thick layer of blubber, has some hair covering. Dugongs are slow, placid animals that live on eel and turtle grasses in shallow tropical waters. They spend their lives in the water, even giving birth there. One puzzling feature of dugong biology is that they have solid bones with no marrow, and it is not known where they manufacture their red blood cells.

Manatees are relatives of the sea cow. They are found mainly in estuaries and in the lower reaches of rivers. Manatees are more social animals than dugongs, congregating in large groups to rest during the day and dispersing to feed at night.

WORLD DISTRIBUTION

■ Dugongs   ■ Manatees

Dugongs are most often found in shallow tropical waters throughout the Indo-Pacific region, with the majority now located in northern Australian waters. The three species of manatees are dispersed along the Atlantic coasts, in the rivers of Africa, and in the Americas between the tropics.

← **Manatees have poor eyesight** and are slow swimmers. This makes them highly vulnerable to human activities, either from deliberate hunting or harassment, or from accidental injury from boat propellers. However, some manatees exploit human use of estuaries by congregating during cold weather in the warm-water outfalls from power plants.

↓ **Dugongs have a distinct rounded head** with small eyes and a large snout. Unlike most other aquatic mammals, dugongs are unable to hold their breath underwater for very long. They can generally do this only for a few minutes at a time, especially if they are swimming quickly.

↓ **Dugongs are markedly different** from manatees because they are truly marine creatures. They also have a distinct trunklike upper lip, with males and old females often having small tusks. The lip is used to pluck up the vegetation from shallow sea-grass beds. Given their appearance, it is a little difficult to comprehend how sailors once mistook dugongs for mermaids.

Whales, dolphins and porpoises

Whales, dolphins and porpoises are the largest group of marine mammals; they have a completely aquatic lifestyle. There is some dispute as to the exact number of species, but it is estimated there are between 79 and 90. These fall into two groups: baleen whales, which are filter feeders; and the toothed carnivorous species, which include the smaller whales and all dolphins and porpoises. With the exception of five species of freshwater dolphins, they are all exclusively marine. All whales, dolphins and porpoises have a hairless, streamlined body with flipper forelimbs and a huge horizontally fluked tail to propel them through the water.

One of the paradoxes of whale biology is that the largest whales eat the smallest food. The largest animals that have ever lived on Earth are blue whales that can reach 110 feet (33.5 m) in length and weigh 196 tons (178 t), yet they feed only on large zooplankton, such as krill. These filter feeders, or baleen whales, get their name from the hairy plates that hang down from the roofs of their mouths and strain huge volumes of water in each gulp.

↓ **Most smaller toothed whales,** such as these bottlenose dolphins off the coast of Australia, live in large groups known as pods, herds or schools. Within these groups, there are complex social interactions that depend on communication through vocalizations pitched at much lower frequencies than those used in echolocation.

ECHOLOCATION
Toothed whales, including the dolphins and porpoises, have evolved an acoustic underwater echolocation system, allowing them to detect prey and predators.

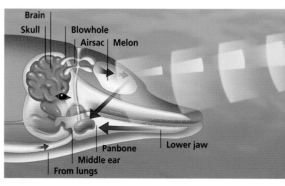

Brain
Skull
Blowhole
Airsac | Melon
Panbone | Lower jaw
Middle ear
From lungs

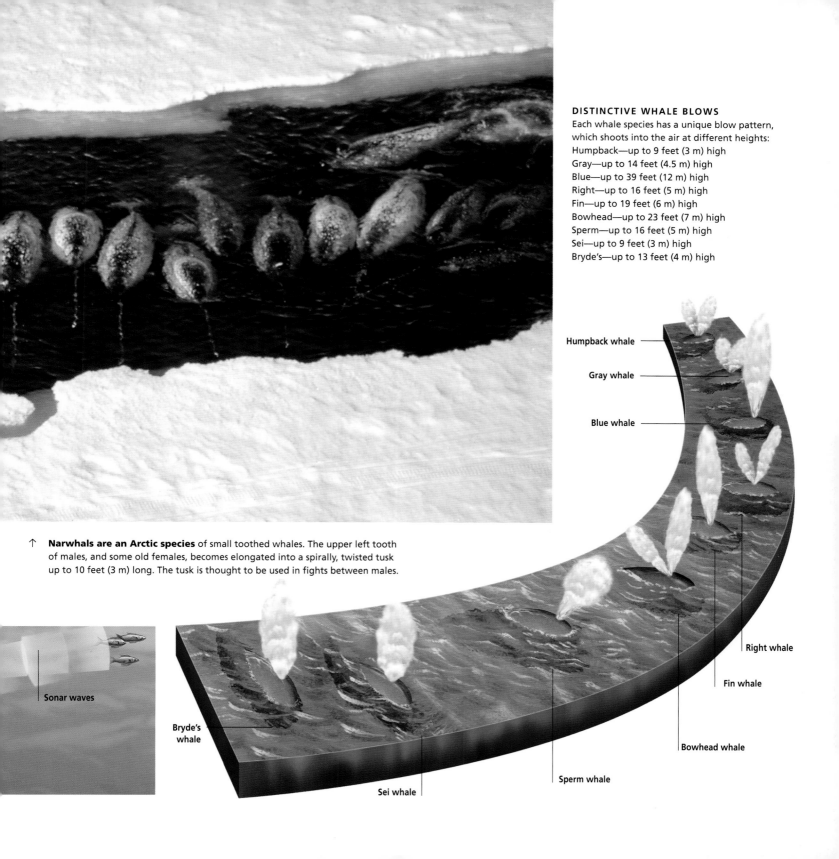

DISTINCTIVE WHALE BLOWS
Each whale species has a unique blow pattern,
which shoots into the air at different heights:
Humpback—up to 9 feet (3 m) high
Gray—up to 14 feet (4.5 m) high
Blue—up to 39 feet (12 m) high
Right—up to 16 feet (5 m) high
Fin—up to 19 feet (6 m) high
Bowhead—up to 23 feet (7 m) high
Sperm—up to 16 feet (5 m) high
Sei—up to 9 feet (3 m) high
Bryde's—up to 13 feet (4 m) high

Humpback whale

Gray whale

Blue whale

Right whale

Fin whale

Bowhead whale

Sperm whale

Sei whale

Bryde's whale

Sonar waves

↑ **Narwhals are an Arctic species** of small toothed whales. The upper left tooth
of males, and some old females, becomes elongated into a spirally, twisted tusk
up to 10 feet (3 m) long. The tusk is thought to be used in fights between males.

Whale migration

Many whale species undertake immense migrations, but the routes of only the gray and humpback whales have been established. The most-studied migration is that of the gray whale along the western seaboard of North America. Gray whales spend May to September feeding in the cold, but food-rich waters of the Bering, Beaufort and East Siberian seas. Once ice forms in these waters, the whales move south. They travel around 110 miles (180 km) per day, either alone or in small groups, until they reach the warm waters of the Gulf of California in mid-December. Pregnant females usually appear first in the south and calve during late February. They remain there until late March, when their calves are strong enough to start the journey northward. Other females mate in these waters, returning 12 months later to give birth.

Migration routes of the humpback whale are less defined. Humpbacks live in both hemispheres, where there are seasonal migrations throughout the year. In both hemispheres, feeding takes place during the polar summers. At the end of the feeding, there is a migration during the winter to breeding grounds in subtropical and tropical latitudes.

WHALE MIGRATION DISTANCES	
Whale	Migration distances (one way)
Bryde's	340 miles (550 km)
Right	1864–2670 miles (3000–4295 km)
Humpback	3500 miles (5635 km)
Minke, fin, blue, sei, sperm	4660 miles (7500 km) (uncertain)
Gray	3726–6210 miles (6000–10,000 km)

A pair of gray whales mates in the warm waters of Baja California, on the Pacific coast of Mexico.

A gray whale surfaces in the winter mating and rearing waters of Baja California, Mexico.

A diving humpback migrates north to over-winter off Tonga, in the warm waters of the Pacific.

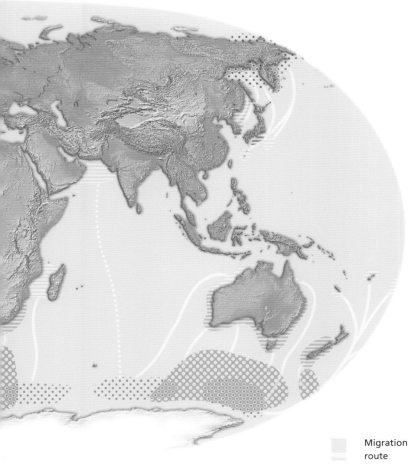

WHALE MIGRATION KEY

Humpback whale breeding area

Humpback whale feeding area

Right whale breeding area

Right whale feeding area

Gray whale breeding area

Gray whale feeding area

Migration route

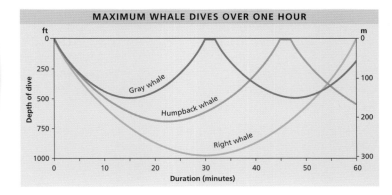

MAXIMUM WHALE DIVES OVER ONE HOUR

Gray whale

Humpback whale

Right whale

Depth of dive — ft: 0, 250, 500, 750, 1000; m: 0, 100, 200, 300

Duration (minutes): 0, 10, 20, 30, 40, 50, 60

DEPTHS OF DIVING

The gray whale can to dive to 500 feet (155 m) and remain submerged for up to 30 minutes. During migrations, it dives only for 3 to 5 minutes. Feeding dives are shallower and are usually less than 18 minutes. Humpback whales have been recorded at depths of 700 feet (210 m), staying submerged for up to 45 minutes. Normally, humpbacks dive for periods between 3 and 9 minutes. Right whales can dive to 985 feet (300 m), remaining submerged for up to 60 minutes. However, dives are usually less than 30 minutes, and only 1 to 10 minutes during migration.

A young humpback leaps out of the water in its summer feeding grounds off Alaska.

A southern right whale over-winters off Argentina. During summer, it returns south to feed.

A southern right whale rolls on its back after breaching. This helps remove dead skin and parasites.

Seals, sealions and walruses

These mammals are known as pinnipeds. They evolved from land carnivores and have become superbly adapted to life in water. Nearly all species are marine, but there are isolated freshwater populations of seals. Although clumsy and ungainly when they haul out, in the water pinnipeds are graceful, agile swimmers. The most obvious adaptations to life in water are their rounded bodies and the modification of their limbs into flippers. Many other adaptations enable them to live in the oceans at all latitudes. Fur and thick blubber insulate them against the cold, their eyes can focus in water, and whiskers acting as feelers supplement their underwater vision. These mammals have a large blood volume and blood chemistry adapted to diving. As a result, they can regularly dive for 20 to 30 minutes down to 330 feet (100 m). The deepest divers are the Weddell seals that are known to dive to nearly 1000 feet (300 m) and remain submerged for 45 minutes. Seals, sealions and walruses have few natural predators, the main ones being leopard seals, large sharks and killer whales. However, hunting, drowning in nets and shooting by fishermen have brought many species to the brink of extinction.

→ **Walruses are bottom feeders,** mainly feeding on large invertebrates. They have a particular fondness for clams. The long, curved, chisel-like tusks are used to dig up buried bivalves, while the whiskers act as feelers in the clouds of sediment that are stirred up.

↓ **The Galapagos sealions** are inquisitive and highly social animals with complex group interactions. The photograph reveals their flexible and streamlined bodies while underwater. They are also extremely agile swimmers—their front flippers are mainly used for swift maneuvers, while the rear flippers provide the bulk of the propulsion.

⤊ **All seals have to come ashore** to breed and rear their pups. Mating is controlled by a small group of dominant males, known as beachmasters, with a single male controlling a harem of up to 50 females on a small stretch of beach. There is inevitably rivalry between males defending their territories or seeking to displace established males, as seen in the photograph above. Adult elephant seals are the largest seals in the world—males can grow up to 16½ feet (5 m) long and weigh as much as 4½ tons (4 t).

↑ **Harbor seals maintain** a stable body temperature of 100°F (38°C), even in extreme conditions. As well as the insulating properties of fur and blubber, heat loss from the flippers is minimized by a heat exchange mechanism that cools blood going to the flippers and transfers the heat to blood returning back to the body. In warmer waters, the flippers act as radiators to cool the blood. Male harbor seals grow to 5½ feet (1.7 m) and females to 5 feet (1.5 m). Harbor seals are common throughout Europe and the east coast of North America.

Seabirds

Only about 3 percent of the estimated 8600 known bird species are described as seabirds, but they are nonetheless found from pole to pole. They are a major component in many marine ecosystems. Seabirds developed from several different groups of land birds, so have widely differing styles of feeding and flight, and endurance capabilities at sea. However, they have common characteristics that define them as seabirds: they rest ashore and return to the land to lay eggs; they have webbed feet used for landing on water and swimming; and they feed on fishes, squid, bottom invertebrates and plankton. All are voracious feeders, as flying and swimming expend a great deal of energy. Seabirds are divided into five groups: penguins, tubenoses, pelicans and cormorants, gulls and terns, and shore birds.

Penguins congregate in vast numbers to breed and raise their young in a colony. The birds constantly shuffle about so that the shifting mass provides some shelter against extreme weather conditions. Penguins usually choose a mate at around four years of age and stay with their partner their entire lives.

Albatrosses are found on all the oceans. They are among the largest flying birds, weighing up to 22 pounds (10 kg). Their huge wingspans can be as much as 11½ feet (3.5 m). Albatrosses are superb and graceful gliders. Gliding is a mode of flight that helps them to conserve energy for their long-distance journeys.

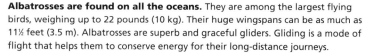

ADAPTATIONS TO THE SEA

Penguins are the most highly adapted seabirds. They have lost the power of aerial flight but effectively fly underwater at high speed, using wings modified into flippers. Thick layers of feathers and body fat insulate them against heat loss in cold water. The tubenoses—albatrosses, shearwaters and petrels—get their name from their distinctive nostrils that contain glands that secrete excess salt, allowing them to drink seawater and remain at sea for long periods. Tubenoses have light bones and, therefore, a light body for their size. Because of their weight, they are able to glide easily. Pelicans, cormorants and frigate birds have strengthened wings and necks so that they can plunge dive or maximize the number of fishes they can carry back from each feeding trip. Shore birds have long legs, light bodies and probing, sensitive bills.

Penguins may be clumsy on land but they are fast, graceful divers underwater. They have been known to reach depths of 2000 feet (610 m) and can remain submerged for up to 20 minutes.

Brown pelicans use their bill pouch to regurgitate large amounts of partially digested fish for their chicks. The chicks grow rapidly, quickly becoming less vulnerable to predators.

Puffins live in shallow burrows on cliff tops, bringing back small fishes to feed to their young. They have the unique ability to hold fishes in their sharp bill while catching additional ones.

Gulls are one of the most abundant inshore seabirds. As natural sources of food have declined or disappeared, they have adapted to scavenging refuse and have moved inland in many areas.

Seabird feeding strategies

Birds from the margin of the seas to the open ocean exploit the oceans as a food resource. Feeding strategies can be divided into various types of food collection, as shown, and into three main zones. The inner zone extends from the shore to a short distance offshore. The main bird types that feed here are gulls, terns, waders, cormorants and pelicans. The offshore zone covers waters out to just beyond the continental shelf edge. The abundance and variety of birds in this zone depend on the width of the continental shelf. The principal species are the gannets, auks and tropic birds. Beyond the shelf break, covering most of the open ocean, the pelagic zone is the realm of the shearwaters, albatrosses and storm petrels.

Birds of the pelagic zone fly long distances to find local pockets of abundant food, such as fishes around plankton blooms. Apart from other rogue birds, birds of this zone are free of the predators that attack terrestrial and inshore birds; they are vulnerable only when they come ashore to breed. The feeding strategy of birds of the pelagic zone has a marked effect on their breeding and life expectancy. The female normally lays a single egg so that the parent is only required to carry food for one chick. Pelagic feeders are long-lived; albatrosses are known to survive for tens of years as adults.

LONG-DISTANCE TRAVELERS

The ability of seabirds to locate fishes in open waters has been exploited by fishermen for centuries; they follow seabirds as a way of homing in on shoals. The distances traveled by many pelagic-zone seabirds are remarkable. One of the greatest seabird travelers is the Arctic tern. It breeds in the Arctic during the northern hemisphere summer, travels 10,000 miles (16,000 km) to feed in the rich waters of the Antarctic during the short southern summer, and then returns north.

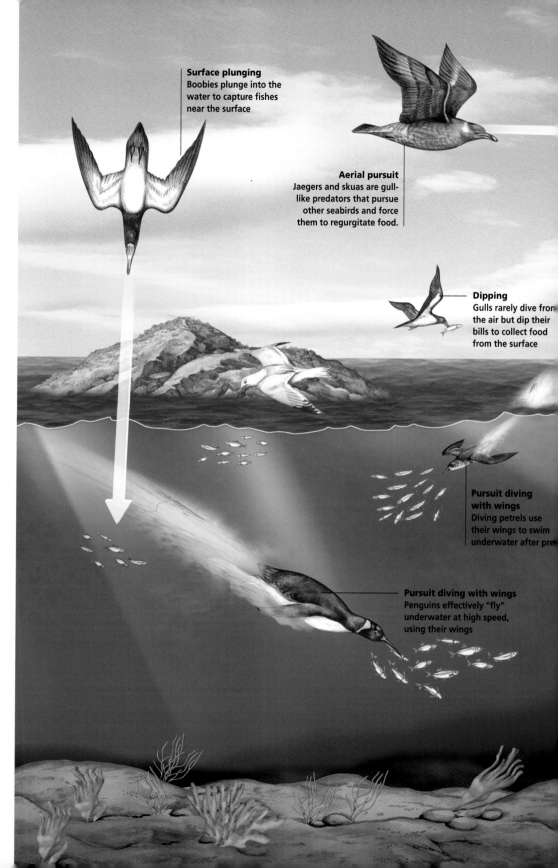

Surface plunging
Boobies plunge into the water to capture fishes near the surface

Aerial pursuit
Jaegers and skuas are gull-like predators that pursue other seabirds and force them to regurgitate food.

Dipping
Gulls rarely dive from the air but dip their bills to collect food from the surface

Pursuit diving with wings
Diving petrels use their wings to swim underwater after prey

Pursuit diving with wings
Penguins effectively "fly" underwater at high speed, using their wings

Surface plunging
Brown pelicans use their pouches to scoop up fishes near the surface

Aerial pursuit
Frigate birds steal fishes from other seabirds and also take fishes from the surface

Pattering
Storm petrels flutter over the waves, their feet touching the water to pick up morsels

Pursuit diving with feet
Cormorants swim with their feet to pursue prey

Pursuit plunging
Shearwaters plunge at high speed, relying on their momentum to carry them underwater

Into the deep

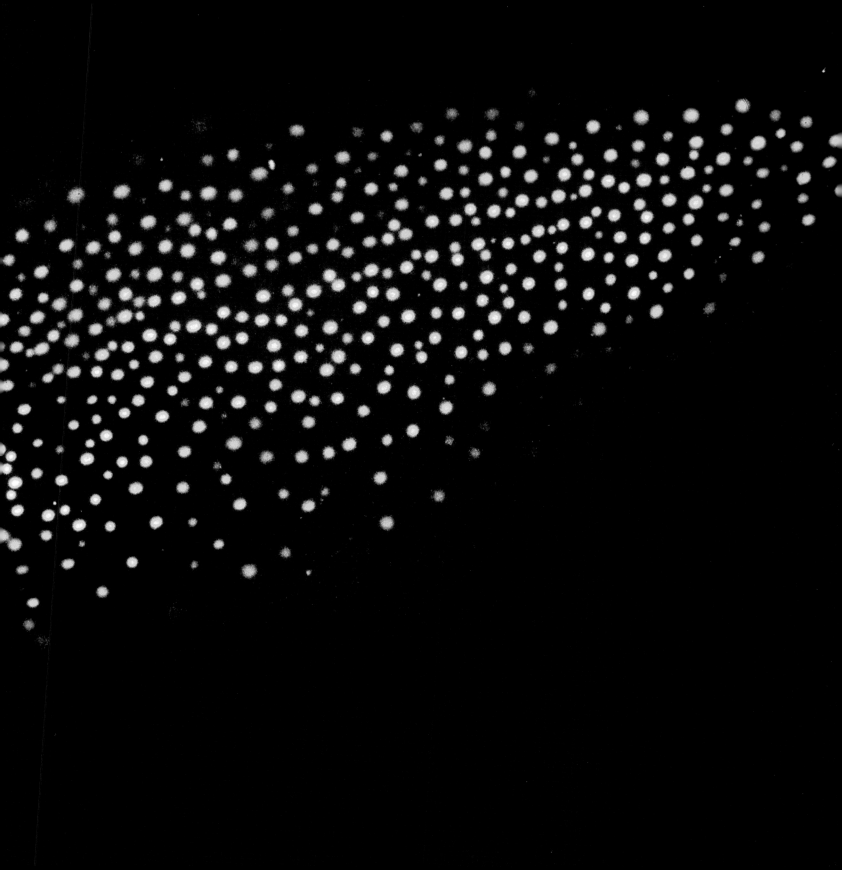

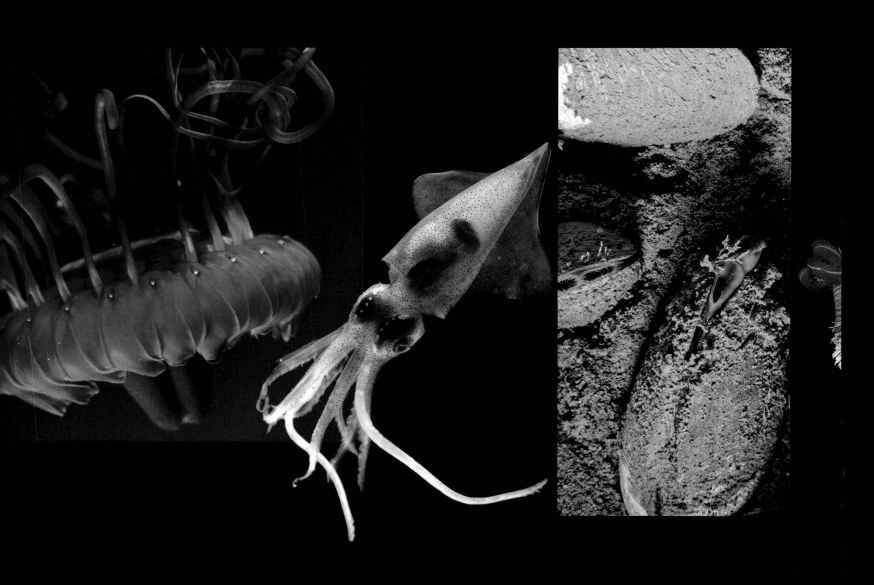

Previous page Many deep sea creatures, like this firefly squid, are able to produce light as a survival strategy.

Into the deep

The depths of the sea are dark and cold. Deep sea life has evolved to deal with the restricted amount of food that reaches the depths from the surface layers. Hydrothermal vent communities are an intriguing exception to the conditions that govern most other ecosystems.

nto the deep

The deep sea begins about 600 feet (200 m) below the surface, where the last vestiges of surface light peter out. Below this, the darkness is broken only by fleeting flashes produced by animals. The temperature regime in the deep sea is quite different to that in the surface layers, with the majority of the deep sea being well below the point at which there is any significant mixing between the surface layers, which are warmed by the sun, and the lower levels. In the deep sea, water temperatures do not vary much from an average of 39°F (4°C). The deep sea is also unique because of the immense water pressures. Most deep sea animals are at the same pressure as their surroundings, so do not have to be physically resistant to the pressure. However, animals that live below 5000 feet (1500 m) display subtle changes to their cell structure and biochemistry in order to cope with the effects of pressure. The deep sea was once thought to be an unchanging place, but research has revealed it to be unstable and subject to sudden cataclysms such as deep benthic storms and immense mud slides that spread over huge distances.

Hydrothermal vent communities were first discovered around the Galapagos Rift in 1977. The large tube worms and other animals were the first known examples of a biological system that was not ultimately dependent on the Sun for energy.

1. Seaweeds can grow only where there is light, so are rarely below 100 feet (30 m).

2. Butterflyfishes feed on coral polyps in waters less than 80 feet (25 m) deep.

3. Anchovy are surface-living fishes, not found below 980 feet (300 m). They feed on plankton and small crustaceans.

4. Bonito are fast- swimming fishes found down to 660 feet (200 m).

5. Marlin are one of the fastest fishes in the open ocean and are found down to 3000 feet (915 m).

6. Jellyfishes float in the surface layers, grazing on planktonic plants and animals.

7. Requiem sharks prey on fishes, turtles and sea mammals and are found to depths of 1150 feet (350 m).

8. Dolphins hunt small fishes, squid and crustaceans. They are found between the surface and 590 feet (180 m).

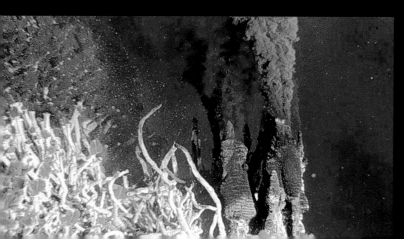

9. Octopuses are stealth predators. Most are found in coastal waters, but a few are found down to 3300 feet (1000 m).

10. Sperm whales can dive to 9850 feet (3000 m) in order to catch one of their preferred foods—large deep sea squid.

11. Lanternfishes live between 980 and 3940 feet (300 to 1200 m) during the day, migrating upward at night to feed.

12. Squid are found from the surface down to the abyssal depths.

13. Hatchetfishes live down to a depth of 5000 feet (1525 m). They migrate upward at night to feed on zooplankton in the surface layers.

14. Deep sea rays feed on the bottom of the sea, eating small mollusks, worms and crustaceans. They are found down to a depth of 9000 feet (2700 m).

15. Viperfishes are found down to 14,440 feet (4400 m). They are ambush predators and feed on passing animals.

16. Grenadiers and rattails feed on dead and live material. They live between 660 and 6600 feet (200–2000 m) down.

...endent on the presence of liquid water. This has resulted in
...re the cradle of life. Most modern hypotheses on the nature
...929, when J.B.S. Haldane, a British physiologist, suggested
... produce organic (carbon-based) molecules when combined
...ng discharges. American scientist Stanley Miller tested the
...experiment producing amino acids—the building blocks of
...arges through a mixture similar to the primitive atmosphere.
...ecules used by living organisms, but the molecules could not
... since suggested that mud or sand in tidal pools may have
...olecules repeatedly replicated themselves. This is a flawed
...rong sunlight needed to energize such reactions was not
...an scientist Jeffrey Bada revisited the problem and suggested
... the site of the origin of life. Clays, or pyrite crystals, close
...uld be more likely to produce self-replicating biomolecules.

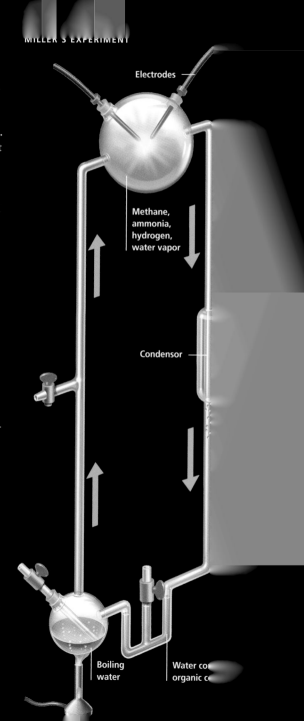

Electrodes

Methane,
ammonia,
hydrogen,
water vapor

Condensor

Boiling
water

Water co...
organic c...

→ **The apparatus set
up by** Stanley Miller
was an attempt to
reproduce what were
thought to be the
atmospheric conditions
on Earth before the
emergence of life. A
mixture of gases rich in
methane and ammonia
was circulated round
the system. Water vapor
was condensed back
to liquid water after
passing through high-
voltage electrical
discharges. Over several
days, the liquid in the
water trap changed
color. It was found to
contain a mixture of
amino acids.

← **For many years,
scientists** believed that
the complex molecules
that became living
systems were first
formed in tidal pools.
It was thought that the
mud provided a surface
on which molecules
would accumulate

ORIGIN OF LIFE IN VENTS

Over the last 50 years, scientists have changed their idea that primitive Earth contained little or no ammonia or methane. It is now believed that the first organic molecules that were able to replicate themselves and become living entities may have formed close to the hydrothermal vents of the deep sea. In 2000, the results of laboratory studies supporting this claim were published. The studies reproduced conditions at hydrothermal vents. They showed it was possible to produce not only amino acids, but also molecules that are key components in the metabolic pathways of living organisms. The combination of iron sulfides and hydrogen sulfide at high temperature, brought together under pressure in the presence of a gold catalyst, produces pyruvic acid. This reacts with ammonia to form amino acids.

↑ **Black smokers contain** deposits rich in iron sulfide and other metals, including gold. Hydrogen sulfide and superheated water flow through the black smokers under tremendous pressure from the surrounding seawater. The combination of various minerals with heat and high pressure is believed to have prompted the development of life on Earth.

→ **In 1953, Stanley Miller** conducted the first significant experiment in tracing the origins of life. Miller duplicated the conditions that would have existed on the early Earth. He combined a "sea" of purified water under an "atmosphere" of hydrogen, methane and ammonia, while electrodes created lightning.

Food, or the lack of it, is the overwhelming biological factor that has quite literally shaped many deep sea organisms. Apart from the exceptions of hydrothermal vents and cold seeps, all food entering the deep sea is derived from the surface of the sea, or has been carried into the deep sea from land. This is because no light can penetrate the depths of the ocean and, therefore, no plants can convert solar energy into plant tissue and provide *in situ* primary production. As a consequence, animals of the deep sea must rely on a rain of plant and animal remains, and a series of overlapping vertical migrations by predators that effectively carry food to the lowest depths—this is known as the ladder of migration. There are losses of energy, as food and material pass through various organisms before reaching the deep seafloor. It is estimated that less than 1 percent of surface primary production reaches the floor of the deep.

MARINE SNOW

In equatorial regions, there is a near-constant supply of food in the deep, albeit at a low level. However, in higher latitudes, the seasons make the supply of food episodic. For example, there are major pulses of plant remains where there are intense spring algal blooms.

Time-lapse cameras left on the ocean floor record the arrival of this material (*right*), the so-called marine snow, and its rapid consumption by the local fauna. These pulses of food act as a trigger to reproduction and other events in animals' lifecycles.

JUNE 22 **JUNE 29**

JULY 14 **AUGUST 10**

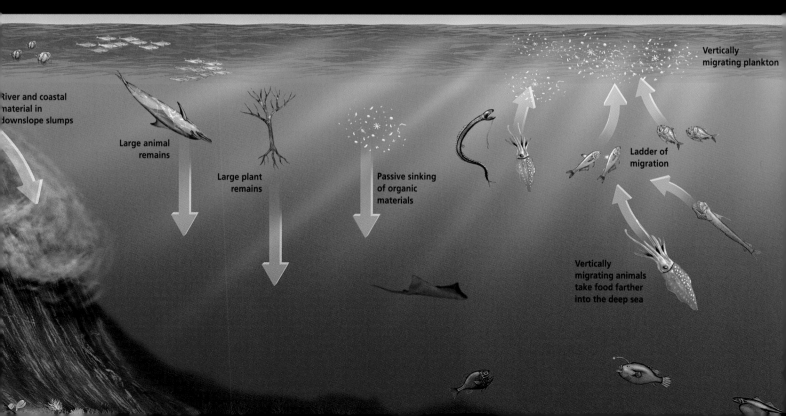

River and coastal material in downslope slumps

Large animal remains

Large plant remains

Passive sinking of organic materials

Vertically migrating plankton

Ladder of migration

Vertically migrating animals take food farther into the deep sea

Once stripped of their flesh by surface scavengers, whale carcasses fall quickly to the seafloor. Studies of whale skeletons on the deep seafloor show that the oily bones can sustain communities of animals, similar to those around hydrocarbon seeps, for many years.

FOOD IN THE DEEP
There are numerous routes by which food enters the deep sea. It is estimated that about 40 billion tons (36 billion t) of organic carbon are formed each year by photosynthetic organisms in the surface layers of the open ocean. However, only 24 percent is still available at 3281 feet (1000 m), and less than 1 percent of the surface production reaches the depths of the abyssal plains at 13,124 feet (4000 m). Large plant and animal remains sink rapidly and carry nutrients normally not present in fine debris. Organic debris is swept into the deep sea from the continental shelf.

Deep sea crustaceans

Many of the large crustaceans common in shallow water, such as crabs and shrimp, are poorly represented in the deep sea. Most are not found below 985 feet (300 m). However, a few of the larger, or megafaunal, species, such as squat lobsters and hermit crabs, are common on the continental slope, yet absent on the abyssal plains. At abyssal depths, there are few species of crabs. Their role as large scavengers is taken over by crustaceans that resemble giant woodlice and can grow to 12 inches (30 cm) long.

In contrast to a rapid decrease in megafaunal crustaceans with depth, smaller, or macrofaunal, crustaceans are extremely abundant. Many biologists call crustaceans the insects of the sea because of their number and the variety of species. The amphipods—the most common of the smaller crustaceans—are scavengers that live in temporary burrows in the surface of the soft sediments covering most of the deep seafloor. An interesting feature of the biology of the macrofaunal crustaceans is their restricted depth and geographic distribution. Many species suppress their planktonic larval stage and produce miniature adults that do not have the capacity to move far from their parents.

↓ **Amphipods are found** in the upper layers of the deep sea. The body is transparent except for a very small amount of reddish pigment in the retina of each of the eyes.

↘ **Stone crabs are one** of the few crab species found in the deep sea. The distinct body shell, or carapace, grows up to 3 inches (7.5 cm) across and the legs to around 6 inches (15 cm) in length.

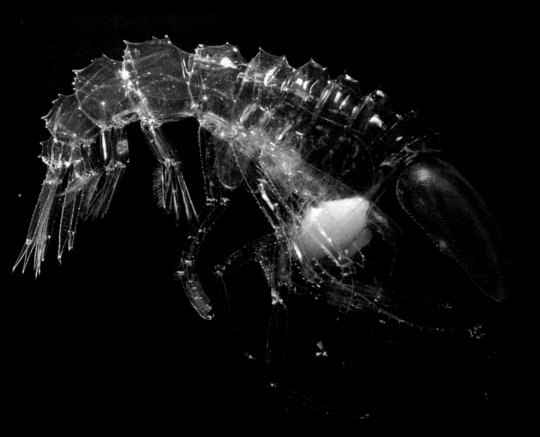

COLONIZATION OF THE DEEP SEA

Whether deep sea animals evolved *in situ* or spread down the continental slope from shallow water is subject to much speculation. Most of the available evidence is based on the abundant and diverse deep sea crustaceans, in particular the amphipods. Deep sea amphipods have body shapes and structures that indicate strong affinities with shallow-water species. From this evidence, it seems likely that there is an ongoing migration of amphipods into the deep sea.

← **The deep sea isopod** crustacean often makes its home inside the bell of a species of deep sea jellyfish. It is thought that this crustacean takes food picked up by the jellyfish's tentacles.

→ **The goblin, or armored, shrimp** is one of the few shrimp species found in the abyssal region of the deep sea, most often at around 19,700 feet (6000 m). These shrimp grow to 3 to 4 inches (7 to 10 cm) in length.

← **Deep sea squat lobsters** live between 3300 feet (1000 m) and 6600 feet (2000 m), where there are around 1000 species. Numbers of species decrease abruptly at a greater depth. At 9900 feet (3000 m), for example, fewer than 20 species are found. The squat lobster exists in great abundance—up to 360 individuals per square yard (330 m²) of one species were recorded in the northeast Atlantic Ocean. Large numbers of squat lobsters also live in the Pacific Ocean, around hydrothermal vents. To feed, they use their delicate front claws to pick up small animals or particles of food from the surface of the sediment.

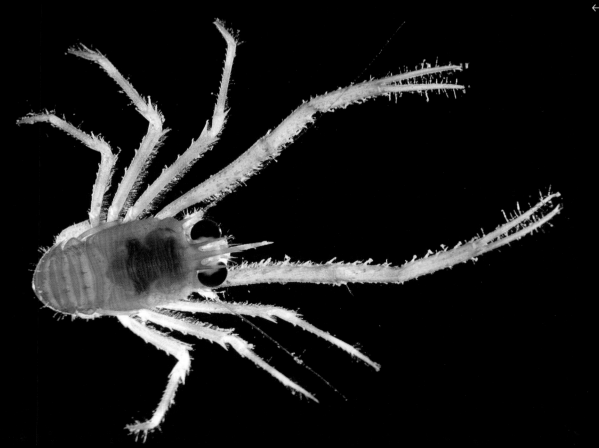

Deep sea squid

For centuries, deep sea squid have inspired seafarers' tales of monsters from the deep dragging ships to their doom. As the farthest depths of the ocean are explored and new species such as the colossal squid (possibly greater than 60 feet/18 m long) are found, these tales seem less fanciful. Squid are cephalopod mollusks that evolved into extremely effective predators. They are found throughout the world's oceans. However, it is in the deep sea that the widest range of adaptations can be seen. Not all deep sea squid are large—tiny cranchiid squid are only 1½ inches (4 cm) long—but all of them are ambush predators. A squid catches prey with a pair of suckered tentacles that it can suddenly shoot out. It then draws the victim toward its mouth; this has a sharp, powerful beak, often covered with venom. To avoid becoming prey themselves, squid use their capacity for jet propulsion to move at high speed. They often emit clouds of ink, which is sometimes bioluminescent, to confuse and irritate predators.

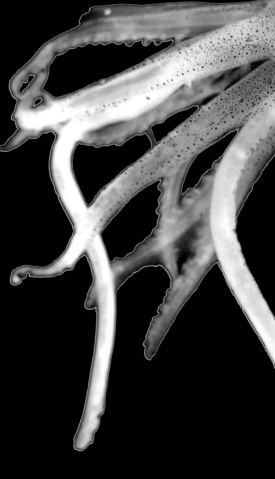

BRAIN POWER

The brain of a squid is among the largest and most complex found in invertebrates. It has evolved to process information, particularly visual information, as well as to exercise a high level of control of the body's nervous system.

A complex brain results in complex behavior. There is evidence of learning and long-term memory in squid. One of the most spectacular manifestations of their advanced nervous system is their ability to control the color and patterning of their skin.

↓ **Most giant squid are known** only from the occasional stranding or from the sucker marks left on sperm whales. The specimen shown here was washed up on a New Zealand beach. The largest specimen was found in 1884 and measured 57 feet (17.4 m) long.

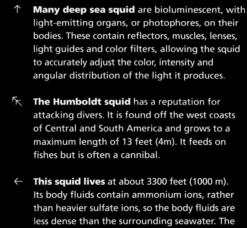

↑ **Many deep sea squid** are bioluminescent, with light-emitting organs, or photophores, on their bodies. These contain reflectors, muscles, lenses, light guides and color filters, allowing the squid to accurately adjust the color, intensity and angular distribution of the light it produces.

↖ **The Humboldt squid** has a reputation for attacking divers. It is found off the west coasts of Central and South America and grows to a maximum length of 13 feet (4m). It feeds on fishes but is often a cannibal.

← **This squid lives** at about 3300 feet (1000 m). Its body fluids contain ammonium ions, rather than heavier sulfate ions, so the body fluids are less dense than the surrounding seawater. The squid can achieve perfect neutral buoyancy.

⇐ **Squid eyes are sensitive** to low levels of light and can detect very small movements. Squid and human eyes are remarkably similar in that both have a focusing lens system, an iris to control the amount of light falling on a highly sensitive retina, and a protective transparent cornea.

Corals, jellyfishes and anemones

Deep sea corals, jellyfishes and sea anemones have the same basic body plan as their relatives in shallow water. However, they have distinctive mechanisms and strategies to cope with the extreme conditions in the deep sea. One of the greatest differences between shallow- and deep-water life is the immense reduction in amounts of food. This paucity affects all types of feeding. Suspension feeders, like the anemones and deep water corals, do not have direct access to phytoplankton and zooplankton. Instead, they rely on non-living food particles dropping from the surface. Inevitably, these lose much of their nutritional value before they reach the deep seafloor. The active predators, such as the medusoid jellyfishes, have far fewer small prey available to them than in shallow water. As a result, deep sea species have low metabolic rates and so can survive on low levels of food. They may grow slowly, but often live longer than a similar species inhabiting shallow water. The complete darkness of the deep sea means that the corals there cannot have photosynthetic zooanthellae as a means of supplementing their food supply. The darkness has also led many species to become bioluminescent.

↓ **Mushroom coral lives** as deep as 4000 feet (1220 m) on hard surfaces in the east Pacific. Like a number of animals in the deep, it grows much larger than its relatives in shallow water, reaching 6 inches (15 cm) in diameter. It is white because it has no zooanthellae. Catsharks use this coral species to protect their egg cases.

← **The spectacular coronate** medusoid jellyfish lives in the bathyal zone of the deep sea, over the continental slope between 656 feet (200 m) to 6560 feet (2000 m). It grows to a maximum recorded diameter of 6 inches (15 cm), and feeds on small swimming crustaceans and organic particles. The vivid red of this jellyfish is a mechanism it uses for camouflage. This species is also bioluminescent.

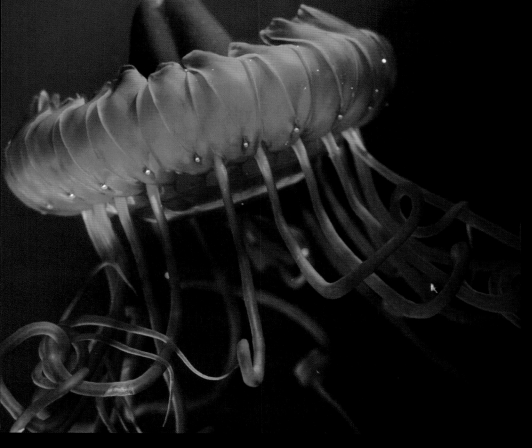

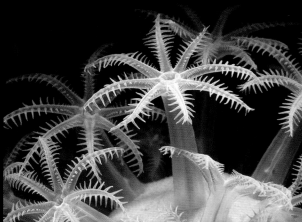

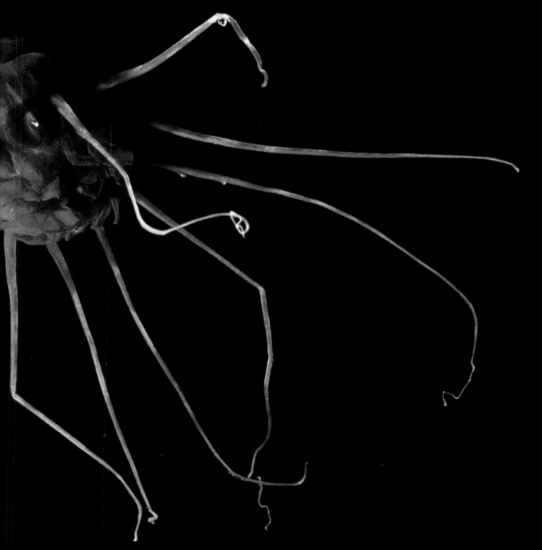

↑ **Siphonophores are colonies** of hydroid medusae that float in midwater. They are very fragile and cannot be sampled with traditional nets. Many species have been observed only since the start of deep sea photography.

← **Deep sea jellyfishes** are active predators, but are also prey for some deep sea fishes. They produce waves of pulsating light on their body surface to attract small prey. They can also deter predators by squirting out a scintillating secretion of thousands of flashing particles.

↓ **Few species of hard corals** live in deep cold water, either in the deep sea or in similar conditions at the bottom of fjords. Cold-water corals grow very slowly, taking centuries to reach any significant size. Hence, repeated damage from deep-water trawling is lethal.

EFFECTS OF THE DEEP SEA

Most comparisons of the body sizes of a wide range of animals in the deep sea with those of their shallow-water relatives reveal a reduction in body size with increasing depth. However, some species of jellyfishes and anemones, as well as some sea spiders and crustaceans, do not demonstrate this relationship between size and depth. Instead, they have become significantly larger than their shallow-water counterparts—a phenomenon known as deep sea gigantism. The causes of this are unclear. However, some clues are provided by the existence of a similar phenomenon in the shallow but cold, stable waters beneath the ice of the polar oceans. It is suggested that when environmental conditions are near-constant for long periods, larger animals are more efficient at food gathering and have an advantage. Smaller animals are usually better able to respond to variable conditions.

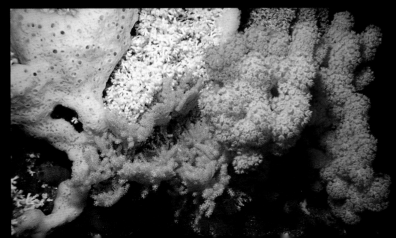

Deep sea fishes

The most abundant fish species and vertebrates on Earth are not one of the shallow-water species such as the sardines or herrings that are found in enormous shoals, but a small mesopelagic species of bristlemouth. Bristlemouths and lanternfishes account for more than 90 percent of all fishes collected in deep sea midwater trawls. Despite their often strange and fearsome appearance, most deep sea fishes are quite small, usually reaching only 1 to 4 inches (2.5 to 10 cm) in length, with thin, often flabby bodies. Exceptions include the lancet fish, which is found in the mesopelagic zone of the deep sea. It can reach the relatively enormous size of 6½ feet (2 m) in length. The benthopelagic black scabbard fish, or black espada, can reach 3⅓ feet (1 m) in length.

Most deep sea fishes belong to primitive groups such as the sharks, eels and less advanced bony fishes. Cuskeels are the deepest living fishes collected by traditional sampling methods at 27,460 feet (8370 m). However, a flatfish, yet to be identified, was seen in 1960 from the submersible *Trieste*, at the bottom of the Mariana Trench, 35,802 feet (10912 m) down.

NOISES IN THE DEEP

Although fishes do not possess vocal cords, a number make noises to communicate. They use sound either as a warning or to attract mates. The deep sea fishes known to make sounds are the grenadier fishes, cuskeels and deep sea cods. The deep sea cods have specialized muscles that drum on the taut, outer surface of their gas bladders. In a number of species, only the males have drumming muscles, but both sexes have the "hearing" part of the inner ear enlarged. Toadfish males call to females over long distances and the deep sea gobies hold courtship conversations. Sound production is also thought to define territories and act as a defence. For example, pony fishes will emit a loud grunt when held.

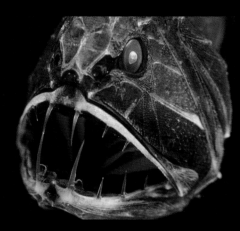

↑ **Fangtooths are found** throughout the world's oceans down to 6560 feet (2000 m) and reach a length of 6½ inches (16 cm). They eat crustaceans and small fishes. The juveniles look so different from the adults, they were once thought to be a separate species.

← **Chimaeras, or rabbitfishes,** have a cartilaginous skeleton and are classified as relatives of sharks and rays. They live on the continental shelf edge and slope, some as deep as 8500 feet (2600 m). They use their blunt snouts to dig up invertebrates and small fishes.

→ **Most barbeled dragonfishes** have no scales, their skin resembling a dark velvet, sometimes with small photophores. The barbel beneath their chin is an effective lure. The gullet and stomach are lined with black pigment so that bioluminescent prey will not shine through the skin, attracting bigger predators.

← **Bristlemouths are the most** numerous fishes in the sea. They range in size from ¾ inch to 12 inches (2 to 30 cm). This species is mesopelagic and lives between 330 to 2300 feet (100 to 700 m), feeding on zooplankton. Bristlemouths exhibit a marked stratification of size with depth, the largest animals living at the greatest depths.

↓ **Mesopelagic fishes, such as the hatchetfishes,** live between 500 and 4000 feet (150 and 1200 m). Here, in very clear waters, there is still some faint light that has filtered down from the surface. This area is known as the twilight zone. Fishes living in this zone have eyes that are up to 30 times more sensitive than those of humans; they are able to detect light where humans can see none.

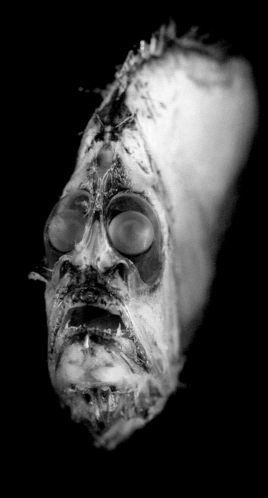

Deep sea worms

Deep sea worms range in size from the microscopic to 6 feet (2 m) long and are one of the most abundant and diverse groups of invertebrates living in the deep sea. Most worms live buried, or partially buried, in the soft sediments that cover the majority of the deep seafloor. Segmented (polychaete) worms are small, but numerous, both in terms of individual numbers and numbers of species. Most polychaete worms feed on organic particles captured with elaborate tentacles, or extracted from the mud as they browse through it. A small number of polychaetes live in, or on, larger animals such as sea cucumbers. Vent tube worms are relatives of the polychaetes; they are much larger and have developed specialized feeding mechanisms to use the chemical energy in hydrothermal vents. Peanut worms (sipunculans) and spoon worms (echiurans) are some of the biggest worms found in the deep sea. They are important in breaking down larger pieces of organic material on the seafloor.

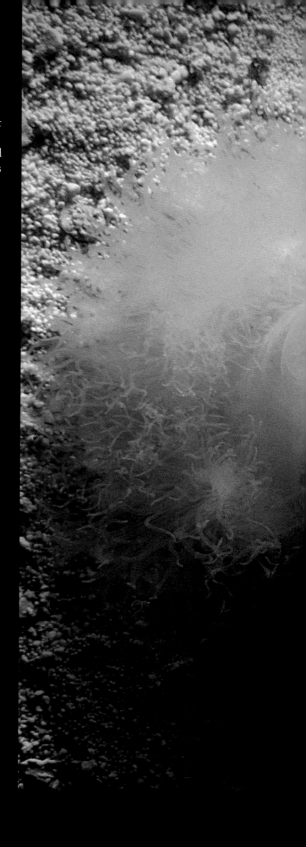

ABUNDANT LIFE
The most abundant group of worms are the non-parasitic thread worms (nematodes). There are thought to be between 10 to 100 million species of nematodes in the oceans. Although less than ⅕ inch (1 mm) long, they can represent 50 to 90 percent of the live weight of animals in deep sea sediments. The diversity of the nematode species means that a sediment sample in one area will be completely different from a sample in an area less than 3280 feet (1 km) away. The sheer abundance of nematodes is such that if all the substance of Earth was removed, leaving just nematodes, the outlines of the ocean basins and continents would be still visible as a mass of tiny worms.

← **The brilliant red gill plumes** on the top of hydrothermal-vent tube worms contain large amounts of hemoglobin. Like the hemoglobin in human blood, the pigment in the plumes is used to absorb oxygen. Gill plumes also absorb sulfides used by symbiotic bacteria living within the worm.

→ **Most deep sea worms** never leave the seafloor. This polychaete worm is unusual because it can swim up from the bottom by waving the mass of tentacles around its mouth.

→ **Not all polychaete worms** live and feed on the deep seafloor. There are between 50 and 60 species that live their entire lives in the water column. The worms are all voracious carnivores, preying on any animal smaller than themselves. They move through the water by using the paddlelike structures on each side of their body.

→ **Tube worms settle in great numbers** around hydrothermal vents. Instead of relying on the Sun, the symbiotic bacteria that live inside vent worms gain energy by converting chemicals ejected from the vents. In turn, these bacteria provide food for the vent worms to live on.

Deep sea cucumbers

Deep sea cucumbers (holothurians) are one of the most abundant groups of megafaunal animals—those that are larger than 0.1 inch (2.5 mm)—living on the deep seafloor. It is estimated that they form 95 percent of the total biomass (the weight of live animals) over large parts of the deep seafloor. Their relative abundance increases with depth. At hadal depths in trench systems, such as the Kuril–Kamchatka Trench, they form 98 percent of all the individuals in trawl samples. Deep sea cucumbers are nearly all surface feeders, continually moving over the soft mud of the deep seafloor, ingesting the veneer of fine particles that have settled out from the water. Some species have been observed in large "herds" and have been likened to bison grazing over the prairies. Not all holothurians are slow-moving surface feeders. Some bury themselves with just the ends of their bodies at the surface, and others live buried headfirst in the mud. A small number, however, have become buoyant and are able to swim off the bottom.

SWIMMING CUCUMBERS

Swimming sea cucumbers can be found in shallow waters, but the accepted view of deep sea cucumbers was that they were long, cylindrical and lived at the bottom of the sea. With the increasing availability of underwater cameras and vehicles, it is has now been determined that many deep sea cucumbers do not conform to this idea. Some resemble flatfishes and swim with an undulating motion just off the bottom. Many deep sea biologists now believe that the flatfish seen during the *Trieste*'s record-breaking dive in 1960 was, in fact, a sea cucumber. To date, about 20 species of deep sea cucumbers have been sighted swimming or floating off the bottom, and some have even been found high up in the water column. One species of deep sea cucumber is now known to spend its entire life in the water column.

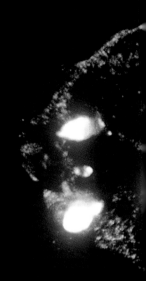

← **This deep sea cucumber** is found between 1970 and 3940 feet (600 and 1200 m) in both the Pacific and Atlantic oceans. It is about 3 inches (7.5 cm) long, and is usually found in dense herds that move over the surface of the seafloor. The leglike protrusions, or podia, on its lower side, keep most of the body clear of the soft mud as it walks, leaving its characteristic track.

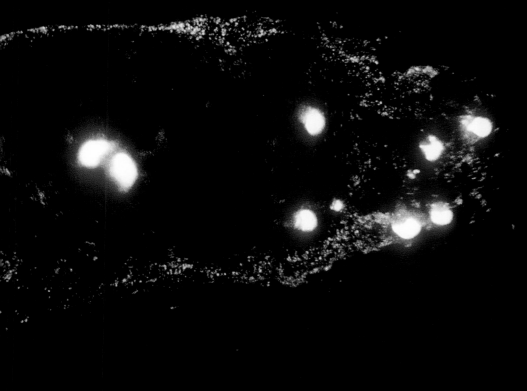

← **This species of sea cucumber** was photographed illuminated only by light it produces in bioluminescent organs, or photophores. This specimen was collected from a depth of 3940 feet (1200 m). It emits blue light when handled but, as with many deep sea animals, it is not possible to determine the purpose of this display. This is because there are few observations of them, as studies of live animals at the surface are not feasible. However, stimulation of bioluminescence by handling suggests that the light display may be used to startle predatory fishes that are sensitive to blue light.

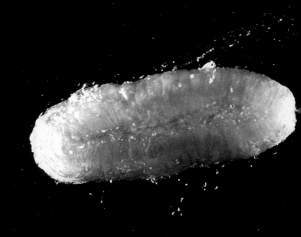

↑ **This species of sea cucumber** was found at a depth of 4920 feet (1500 m). Its feeding activities have been observed as part of a long-term study in the northeast Atlantic. Here, there is seasonal deposition of organic debris following the Spring Bloom in surface waters. This material is eaten by holothurians, while organic material in their fecal cast is devoured within an hour by deep sea urchins.

← **It is thought that deep sea** holothurians can process up to 3½ ounces (100 g) of sediments per day, though it is believed that most species can also absorb nutrients directly from the water. This makes holothurians an important component in the ecology of the deep sea. They control the populations of smaller buried animals, resuspend sediments and return dissolved nutrients back to the water.

The floor of the deep sea occupies 58⅓ million square miles (151 million km²) of Earth's surface. This represents 41 percent of the world's oceans and 29.5 percent of the planet's surface. The majority of this enormous expanse of seafloor is covered in soft, fine sediments accumulated over millions of years. Deposits on the deep seafloor consist of countless billions of calcareous and silica shells of microscopic organisms, as well as dust particles blown from the land. There are even tiny meteorite fragments, known as microtekites, in deep sea sediments. The conditions in the deep sea have been summarized as deep, dark, cold and short of food. As a consequence, the main biological characteristic of life on the seafloor is the lack of it. For example, on nearshore seafloors, there are approximately 1½ pounds of organisms per square foot (5 kg/m²). On offshore sediments on the continental shelf, there is approximately 1 ounce of organisms per square foot (200 g/m²). In the deep sea, there is less than $\frac{1}{12,000,000}$ ounce of organisms per square foot (1 mg/m²). The restricted food supply and low temperatures cause deep sea organisms to grow slowly. However, they are often long-lived. The relative stability of the conditions may result in some species becoming "giants." The main groups of animals living on the surface of the deep seafloor are the coelenterates, brittlestars, sea cucumbers, crustaceans and fishes.

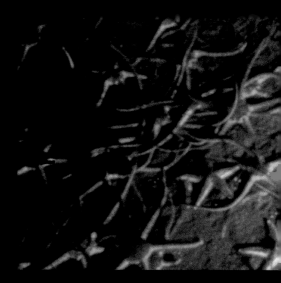

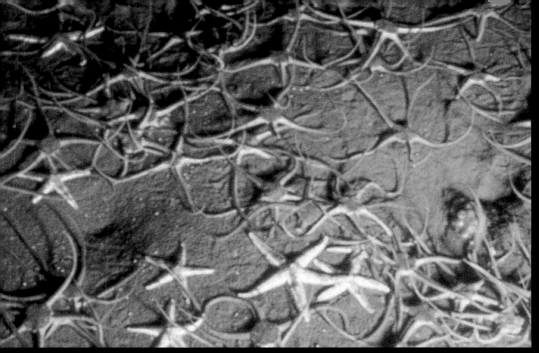

SEAFLOOR ADAPTATIONS

The soft surface of deep sea sediments makes it difficult for large surface dwellers to move across without sinking or using up valuable energy as they attempt to move forward. The need to conserve energy for growth and reproduction is a strong evolutionary pressure in a food-poor environment. Some sea cucumbers that feed by ingesting the surface layer of the sediment and digesting organic material have become slightly buoyant so that they need less energy to move over soft mud. One species has a sail-like tail filled with fluid that is less dense than seawater. Its back is raised and only the front end, the mouth, pushes into the sediment.

↑ **Brittlestars may be** the most widely distributed and numerous of the larger deep seafloor surface-dwellers. They are found at nearly all latitudes and often in the arm-tip to arm-tip densities seen here. Their low metabolic requirements make them well suited to the slow pace of life in the deep sea.

← **Ratfishes are mainly found** on the continental slope, where there is a little more animal life in the surface layers of the seafloor, rather than on the abyssal plains. Acute olfactory and electrical senses enable ratfishes to locate buried bivalves and worms.

→ **The pom-pom anemone** can grow to 10 inches (25 cm) across. It takes on a variety of shapes—from low and flat to nearly spherical—depending on the strength of the weak, near-bottom currents. Unlike anemones that live on hard surfaces, the pom-pom anemone cannot move by flexing its pedal disk. It has been observed, however, rolling like tumbleweed over the seafloor, moved by the current.

Adaptations: sense organs

It is a paradox that in the darkness of the deep sea, many animals have evolved extremely complex and highly sensitive eyes. These help to discern the faintest glimmer of surface light and occasional flashes of bioluminescence. Many deep sea animals also have highly developed olfactory senses for detecting food or mates in the deserted expanse of the deep sea. They can detect chemicals leaching from food items or pheromones from potential mates. The senses of touch and hearing are clearly separate in terrestrial vertebrates like humans, but this distinction is less clear in the deep. Here, water is a far better transmitter of low-frequency pressure waves than air. What we regard as the sense of hearing is, in many deep sea animals, a long-distance sense of touch—they detect vibrations from other animals in the deep. Many invertebrates hear/touch using sensitive hairs or antennae, and in fishes the "lateral-line" system functions like our inner ear—sensitive hairs transform microscopic movements into nerve impulses. Many fishes produce sound so they must also possess a sense of hearing as we understand it. Finally, most active deep sea animals are able to detect gravity. Invertebrates have simple receptors called statocysts, but fishes have more elaborate semicircular canals. These not only provide information on vertical orientation but can also assess changes in speed.

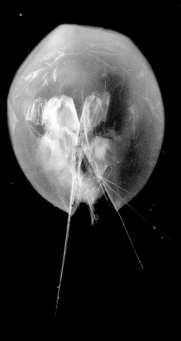

→ **Deep sea ostracod crustaceans** are believed to have the most light-sensitive eyes of any animal. Each eye consists of a pair of forward-facing parabolic mirrors. A group of light receptor cells hangs like a light bulb close to the focus of each mirror. The image may not be clear but this system does collect and detect any light.

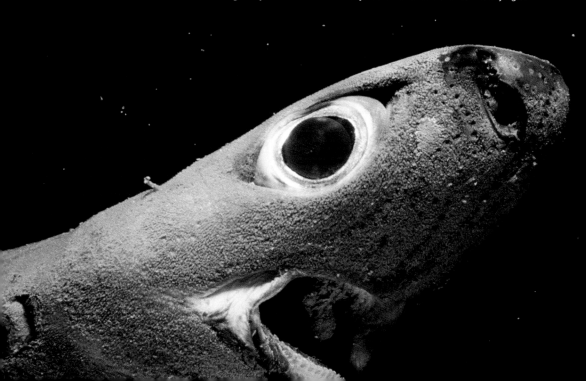

← **The bioluminescent velvet belly** lanternshark has much larger eyes than its shallow-water relatives. This ensures it can gather what little light is present in the mesopelagic zone. It also has much larger nostrils so it can sample more water with its highly developed sense of smell. It can detect low concentrations of chemicals emitted by other animals, using convoluted nasal passages lined with millions of olfactory cells connected to a center in the brain. It also has a highly developed electrical sense that can detect bioelectric fields and Earth's magnetic field.

VISUAL ADAPTATION TO THE DEEP

Midwater fishes in the deep sea have large and exquisitely sensitive eyes. Similarly, sensitive eyes are also found in squid, shrimp and other invertebrates. To improve on the visual efficiency of the "normal" fish eye, many species have developed tubular eyes. The large lens and multilayered retina of a deep sea fish eye improves the gathering and detection of all the available light in a given direction. As a further refinement, some species have secondary lateral lenses and extended retinas to improve their lateral vision. Some also have yellow filters that distinguish between ambient light and bioluminescence.

NORMAL FISH EYE

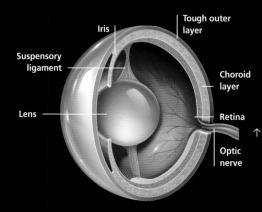

- Iris
- Tough outer layer
- Suspensory ligament
- Choroid layer
- Lens
- Retina
- Optic nerve

TUBULAR FISH EYE

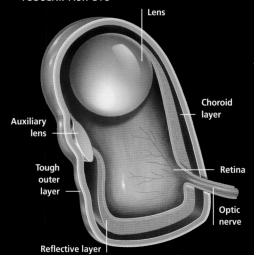

- Lens
- Choroid layer
- Auxiliary lens
- Tough outer layer
- Retina
- Optic nerve
- Reflective layer

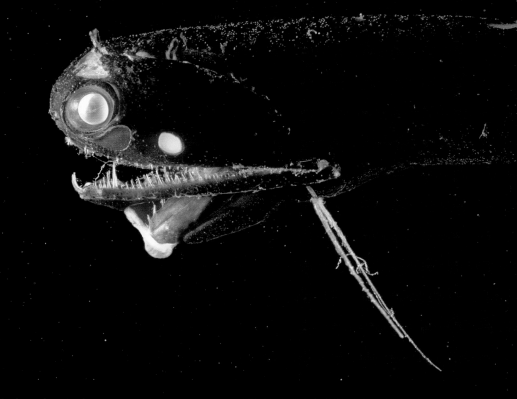

↑ **The large lens in the eye** of this mesopelagic fish is able to gather light from a wide visual field. It focuses light onto a single point on the retina that is especially sensitive, mainly because of the high concentration of retinal rod cells.

→ **The hairy anglerfish** has modified its fin rays into sense organs. The rays are sensitive to vibrations in the water and supplement the information from the fish's highly effective lateral-line system. They are also covered in chemosensory cells, so that the fish is able to taste/smell the distant presence of other animals.

Adaptations: color

The types of coloration of deep sea animals are a response to the need for camouflage. This is a means of disguise for predators lying in wait, or for avoiding becoming prey. The underlying function of the various colors and patterns is either to produce some form of countershading, or to blend with the background by becoming semitransparent. There is often some residual ambient light in the upper mesopelagic zone of the deep sea. This is where jellyfishes, shrimp and bristlemouth fishes are found in varying degrees of transparency. Coloration changes markedly in the deeper parts of the mesopelagic zone. Fishes become silvery as a form of countershading, or, at the lowest levels of this zone, velvety black to absorb what little light may be present. Invertebrates in this zone are typically orange or red, which may be partly a consequence of a diet rich in red and orange pigments. Red light is completely absent at these depths. Hence, the invertebrates appear black or an ill-defined gray when illuminated by the feeble blue light that penetrates this far down. There is no light in the deepest part of the ocean other than bioluminescence, and no consistent trend in coloration. Most animals in the deepest seas lack any strong coloration.

↑ **This spectacular comb jelly** is found in the lower mesopelagic waters of the Pacific, where its red coloration renders it almost invisible. A voracious predator of other comb jellies, it uses its camouflage to avoid both detection and being eaten by midwater fishes. This vibrant comb jelly quickly loses its color if kept in bright light.

← **These shrimp live** in the lower portion of the mesopelagic zone of the deep sea. Their bright red coloration is nearly invisible in the extremely faint blue light that dominates this far down. Any light that the shrimp reflect will not be detected, as most animals at this depth are insensitive to the color red.

↓ **The spiny lanternfish** lives in deep sea between 650 and 3300 feet (200 and 1000 m). At these depths, a reflective silvery coloration breaks up the fish's body outline, ensuring it is a difficult meal to detect. The silver coloration is produced by mirrorlike plates of guanine crystals, found beneath the skin.

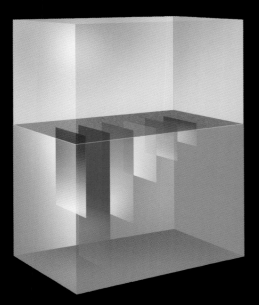

COLOR AND DEPTH

White light is composed of a spectrum of wavelengths that we see as colors. When light enters the water, the red end of the spectrum (longer wavelengths) is quickly absorbed. The shorter green and blue wavelengths can travel farther. The absence of red gives underwater light its blue coloration. All the wavelengths are eventually absorbed. So, even in the clearest seawater, the light can penetrate only to a maximum depth of 4265 feet (1300 m).

↑ **In the upper layers** of the deep sea, there is sufficient light for animals to be detected by their silhouette. To minimize detection, a number of species have modified the proteins in their bodies to have the same optical density as water, and so become effectively transparent. However, eye pigments cannot be made transparent and are clearly visible.

← **Anglerfishes live in the complete** darkness of the bathypelagic zone. The fish's dark, scaleless, velvety skin, absorbs any light, and helps the fish blend in with the surrounding darkness. This is crucial to this species because it must avoid being illuminated by the light emitted from its own bioluminescent lure.

Adaptations: shape

The fishes are one of the main groups of deep sea animals that display some of the greatest differences in body shape when compared to their shallow-water relatives. This is probably a consequence of the scarcity of food in the deep sea and the various strategies fishes have evolved to cope with it. Active, sustained pursuit is energetically expensive in such a food-poor environment, so most fishes are ambush predators and spend most of their time hanging motionless in the water. This means their bodies do not have to be hydrodynamically efficient. They can be either deep and thin, or long and thin, with a reduced musculature and skeleton. The flesh of deep sea fishes is flabby and watery in comparison to that of shallow-water species; this is due to a lack of muscle fibers. Deep sea fishes all have large mouths and sharp teeth to ensure that, on the rare occasions when prey is encountered, it does not escape. It is thought that the relative extreme length of some species is a means of increasing the length of the lateral line—a series of sensory organs used by fishes to detect prey. Some fishes, such as anglerfishes, are globular rather than thin, and are weak swimmers because they lure their prey right into their mouths.

↓ **The gulper eel lives in the eastern Pacific** at 7800 feet (2400 m) below the ocean surface. It reaches a maximum length of 6 feet (2 m) from the head to the tip of the slender tail. The gulper eel is little more than a huge set of distinctive jaws attached to a stomach, having no scales, ribs or gas bladder.

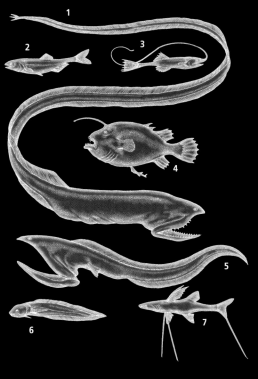

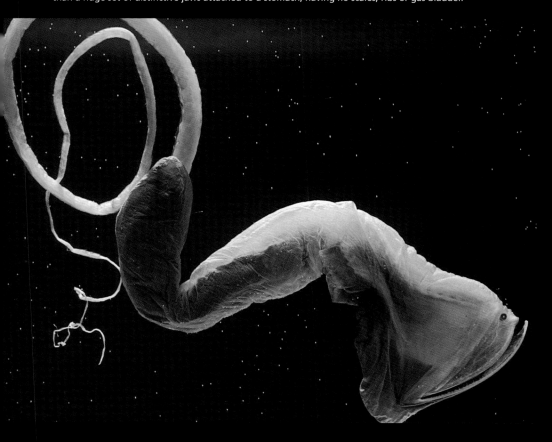

BODY SHAPES
1. For a deep sea fish, the swallower is large at up to 63 inches (1.6 m) long. **2.** Deep sea bristlemouths have weak muscles compared to their relatives in shallower waters ; this is because they spend most of their time hanging motionless in midwater.
3. In the whipnose, females are around three times the size of males and can reach lengths of 14 inches (35 cm), excluding the lure. The long lure can be up to four and a half times the body length. **4.** In some species of anglerfishes, a dwarf male is permanently attached to a larger female. **5.** The jaws of the pelican eel extend backward, enabling the eel to have an enormous gape. **6.** The deepest fish found, the cusk eel, has a distinctive snout with soft and eel-like fins. **7.** Tripodfishes have long extensions of their pelvic and tail fin-rays that they use as supports to hold them above the seafloor.

→ **Hatchetfishes live in the mesopelagic** zone between 1640 feet (500 m) and 4900 feet (1500 m), where there is often a small amount of ambient light. This fish's body is compressed laterally so that its outline is as small as possible when viewed from above or below. The outline is further camouflaged by the mirrorlike crystals in the skin and lines of photophores along the body. The stomach and jaws of this species are enormous relative to the rest of the body.

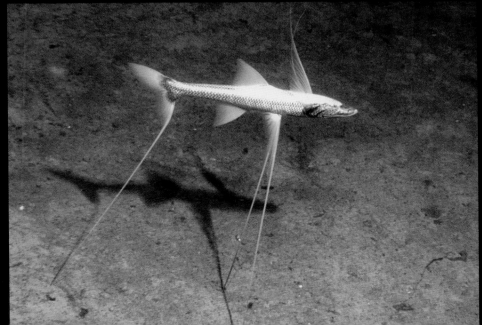

BIG MOUTHS

One of the most consistent characteristics in the body shape of deep sea animals is a relatively large mouth. Comparisons of species at different depths have shown a relationship between depth and mouth size—the deeper the animal, the larger the mouth relative to body size. The deep dwellers also have larger heads to accommodate well-developed gill rakers—the bony or cartilaginous projections which prevent small prey from escaping their mouth. The larger head and mouth allow them to extend their prey size range in both directions. Since encounters are rare, all deep sea fishes are assumed not to be too selective in the size and type of prey.

← **Tripodfishes "stand" motionless** for long periods, waiting for small crustaceans and fishes to pass. By remaining immobile, they reduce background noise and increase the sensitivity of their senses.

Adaptations: bioluminescence

Many organisms are bioluminescent; that is, they can produce light without heat by the action of the enzyme luciferase on the substance luciferin in the presence of oxygen. In the deep sea, bioluminescence is used by animals to provide counter-illumination to break up their silhouette, to lure prey, to distract predators and to signal other members of their own species. Some fish species are able to produce red light. This is invisible to most deep sea species and enables the predator to illuminate potential prey without alerting them to their presence. In animals such as jellyfishes and comb jellies, light is produced in individual cells spread over the animal's surface. In most cases, however, light is produced by specialized and often complex organs called photophores. In some species, the luciferase and luciferin are manufactured by the animal itself. In others, the photophore contains bioluminescent symbiotic bacteria nourished by the host. Photophores are used to control the color, intensity, direction and flashing rate of the emitted light. However, not all bioluminescence is confined to cells or photophores. Many species have not only photophores, but also specialized glands to squirt out bioluminescent fluids to confuse and divert predators.

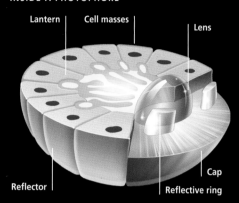

INSIDE A PHOTOPHORE

Lantern | Cell masses | Lens | Cap | Reflective ring | Reflector

PHOTOPHORE CELL

Photophores have certain similarities to an eye but instead of detecting light, these organs produce and carefully regulate the light signals that are emitted. Light without heat is produced by bringing together luciferin and the enzyme luciferase in the cell mass. The light is guided in the lantern toward the lens. There, it is focused into a forward-pointing beam. Any stray light rays are blocked by the reflective ring in the transparent cap. The cup-shaped reflector helps to focus light from all directions onto the lens, maximizing the output. Pigment surrounds the entire photophore to prevent illumination of surrounding tissues. This ensures that the only light visible is that passing through the front of the photophore, where it can be controlled for intensity and direction.

1

2

3

4

↖ **The significance of photophores** is shown in this progression of illustrations. Photophores break up an animal's silhouette, making detection difficult.

← **Firefly squid have photophores** on the underside to break up their silhouette, and on the arms for sexual signaling. Although squid do not have color vision, some species can distinguish between surface light and bioluminescence.

→ **This jellyfish species** produces a bright light from cells scattered over its body and tentacles. The light is thought to attract zooplankton and small fishes.

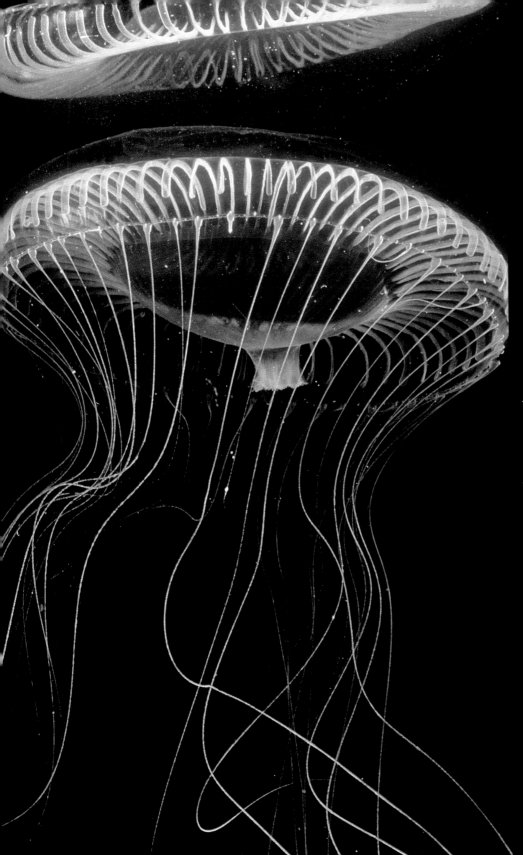

↑ **This deep sea starfish** was photographed lit only by the photophores covering the upper surface of i body. Starfishes feed on small animals and detritus in the surface layers, so the bioluminescence is most likely used to startle predators.

↓ **Flashlight fishes** use light produced by bioluminesc bacteria in pouches beneath each eye and the ligh can be turned on and off by covering the pouches with flaps of skin.

Adaptations: feeding

e restricted food supply in the deep sea has been the driving force behind many extreme aptations in its animal inhabitants. Food reaching the deep sea can be categorized into e following types: large lumps that include live prey and rare carcass falls; the fine rain of aller, less nutritious particles originating in the surface; and dissolved nutrients. The types animals that exploit these different food resources is determined by whether they live in water column (pelagic species) or on, or in, the sediments (benthic species). Most benthic cies rely on the settlement and accumulation of organic particles on the seafloor, feeding this material directly, or on small organisms in the surface layers. Pelagic animals are more datory because fine particles are in very low concentrations in the water column. Eating e prey is a much more efficient form of feeding. The main predatory species—the fishes I squid—most clearly show the predatory adaptations and strategies for deep sea living.

RING PREY

riety of bioluminescent lures is one of the most mon adaptations seen in deep sea predators. s simplest form, a lure is a glowing lump on a k waved in front of the gaping mouth of the dator. In some species of anglerfish, however, lure is far more elaborate. In addition to a lure waves above the mouth, luminous, branched

barbels hang below the chin. Dragonfishes have glowing chin barbels of great length, and some squid have photophore lures on long tentacles. It is reasonable to think that these lures are at risk of being bitten off by prey, but records of such an occurrence are rare. Other species have photophores lining their mouths, enticing prey right inside.

↑ **A dragonfish has a long** bioluminescent lure hanging from its chin. These are the longest of this type of lure seen in deep sea fish. Instead of the lure being illuminated by internal bacteria, the fish generates its own bioluminescence.

↓ **Loosejaw fishes have** very long jaws, resulting in a large gape. These fishes can dislocate their jaws and swing them forward to further increase the gape of the mouth. Large prey are, therefore, a viable target.

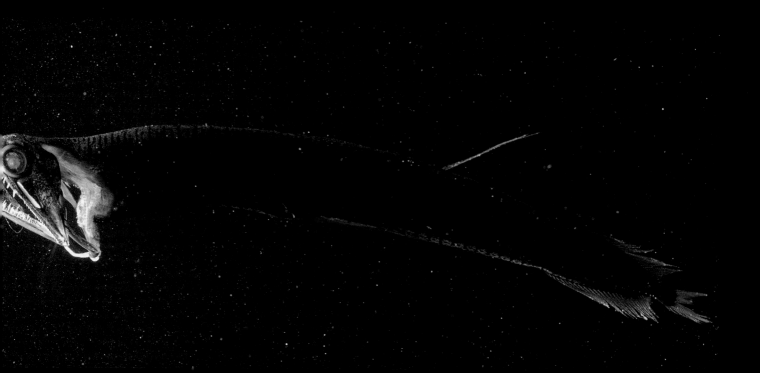

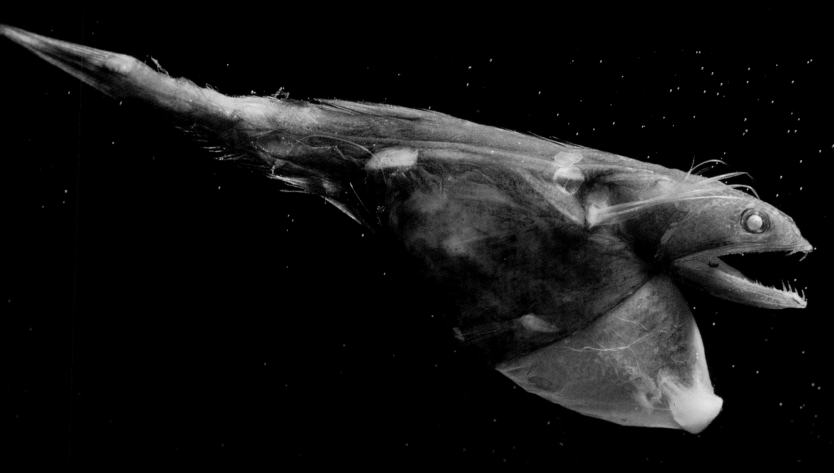

VIPERFISH

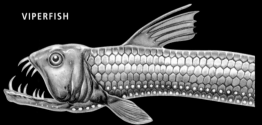

VIPERFISH FEEDING

VIPERFISH JAWS

Many fishes have expansive jaw mechanisms that enable them to swallow animals whole. The upper and lower jaws of the viperfish can swing forward and pivot around a specialized neck joint behind the head, increasing the gape of the jaws. Food can pass between the sharp fangs without being impaled and travel directly into the top of the gullet.

↑ **Black swallowers are small** fishes whose stomachs can greatly expand. These fishes are able to swallow prey larger than their own bodies.

→ **Cranchiid squid are active predators.** Some avoid becoming prey themselves by tucking their head and tentacles into their body. They can then inflate their body into a ball, sometimes filling it with ink.

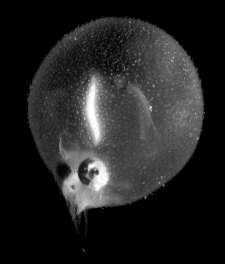

Adaptations: reproduction

Animals of the deep sea live in a vast, sparsely populated environment where encounters with other creatures are infrequent. The difficulties this presents in terms of feeding are overcome by mechanisms that ensure a wide range of food sizes and types can be eaten, but the chances of encountering an individual of the opposite sex and the same species are considerably rarer. Deep sea animals have evolved a number of strategies to cope with this problem. Signaling and attractants, such as sound production, bioluminescent patterns and chemicals known as pheromones, increase the possibility of such encounters. The chances of a fruitful encounter are also enhanced by hermaphrodite individuals or by the forming of long-term pair bonds. This is taken to an extreme form in some anglerfish species. Successful reproduction depends not only on getting parents together, but also on the biology of the offspring. When conditions are stable for long periods, the animals in the lower layers of the deep sea take the risk of investing in a small number of large, yolky eggs. These eggs have a very short larval stage and most of the offspring will survive. In contrast, species in shallow water produce large numbers of small eggs, most of which die.

SEARCHING FOR MATES
One of the biggest problems for deep sea animals is how to find a mate in the vastness of the ocean depths. Some fishes have overcome the problem by becoming hermaphrodites. Another strategy is to hold on to a mate permanently. This has been refined by a number of deep sea anglerfishes. In these species, females grow into full-sized adults but with inactive ovaries. The males, however, are dwarfs by comparison. Pheromones attract a male to a female. He attaches himself to her and eventually their blood vessels merge. Once attached, the gonads (reproductive organs) of both animals are stimulated to ripen, resulting in the production of fertilized eggs. The male is totally reliant on the female and, over time, degenerates into little more than a testis.

↑ **Anglerfish larvae** migrate upward to find more food of a suitable size. The balloonlike envelope of fluid-filled skin around the larva is characteristic of the early stages of anglerfishes. It is thought to make the fish neutrally buoyant, so that no energy is expended in staying suspended in the water column.

← **Some species of female anglerfishes** produce pheromone attractants that draw much smaller males to them. Most females carry only one dwarf, or parasitic male. Females with two males are not uncommon; occasionally three males have been observed on a large female.

← **This young velvet belly lantern shark** has a large yolk sac from which it draws nourishment until it is able to feed itself. This deep sea shark species produces large, yolky eggs that are retained within the female, where they hatch. The female gives birth to well-formed juveniles, which she does not feed as they develop. Instead, they must rely on their initial supply of yolk.

→ **This adult male anglerfish** has large olfactory organs to help detect a female in the vast abyss. The male uses sharp hooks on the snout and chin to attach himself to the female. Males are 5 to10 percent the size of adult females. The females can grow to approximately 18 inches (45 cm) in length.

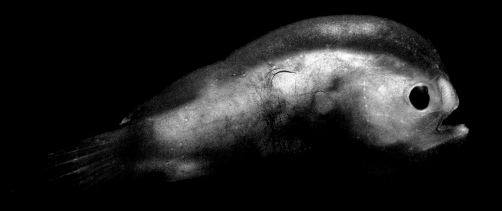

Hot vents

One of the most exciting events in marine science occurred in 1977, with the discovery of hydrothermal, or hot, vents. While the existence of flows of mineral-rich hot water around the ocean ridges had been predicated for years, their actual discovery was startling. During an expedition in the submersible *Alvin* to explore the East Pacific Rise, an ocean ridge near the Galapagos Islands, geologist Robert Ballard and biologist J. "Fred" Grassle came upon huge rock chimneys from which gushed plumes of superheated water resembling black smoke. The two scientists had discovered a black smoker vent. Vents have been found in other parts of the Pacific, the Atlantic and Indian oceans, and even in freshwater on the bottom of Lake Baikal in Siberia. It is the sulfide content in the water that produces the appearance of black smoke. Not all vents are black smokers. Some are cracks in the seabed, or mounds of porous material through which cooler hydrothermal fluids percolate, having mixed with the cold surrounding seawater below the surface.

LIFE WITHOUT THE SUN

The discovery of the animal communities living around vents was of immense scientific importance. It had been thought that all life was ultimately dependent on solar energy carried through food webs, but the anatomy and biochemistry of many of the species showed that these animals were nourished by bacteria within their bodies. The bacteria derived energy from the sulfides in the vent water and were independent of any food produced by photosynthesis at the surface.

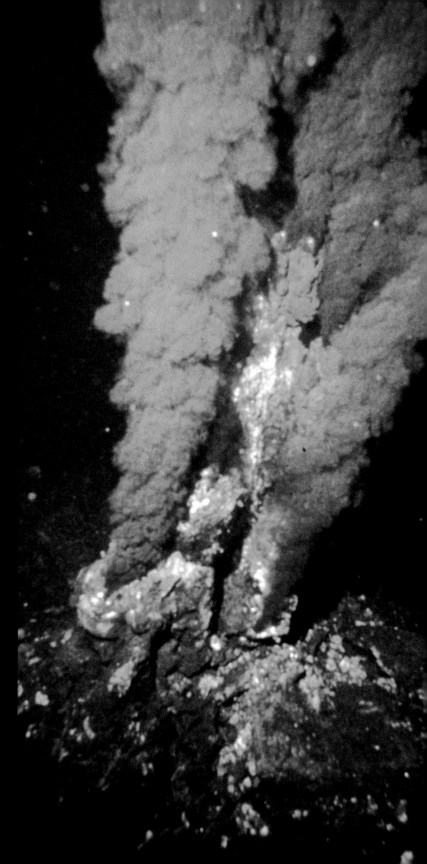

INSIDE A BLACK SMOKER

→ **Seawater forces its way deep** into the crust through cracks around active ridges. Contact with hot rocks heats the water; it then dissolves surrounding minerals and rises to the surface. The water is at 662°F (350°C), but does not become steam as it emerges because of the surrounding water pressure. Minerals are deposited around the main vent. Water also escapes from secondary vents and fissures.

← **Black smokers get their color** from particles containing large amounts of sulfides, lead, cobalt, zinc, copper and silver. The mineral composition of the plumes changes from vent to vent and over time. Some vents are white smokers—these produce streams of water rich in gypsum and zinc but scant in copper or iron.

←← **The Pompeii worm** is a species of polychaete worm adapted to life at hydrothermal vents. It lives close to the hottest water from the vent, and contains proteins modified to withstand high temperatures. The Pompeii worm cultivates sulfur-eating bacteria on immense plumes of feathery gills.

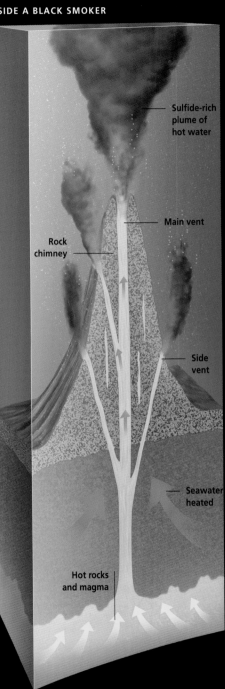

Sulfide-rich plume of hot water

Rock chimney

Main vent

Side vent

Seawater heated

Hot rocks and magma

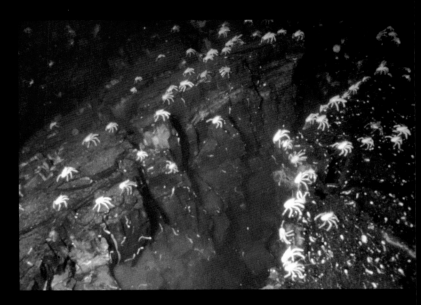

↑ **Not all vents are large chimneys** surrounded by giant tube worms. The water from this 3-feet (90 cm) wide crack in the East Pacific Rise contains sulfides and other minerals that feed a mat of bacteria. In turn, the bacteria is grazed on by innumerable white crabs and other animals living around the fissure.

↓ **Tube worms living around vents** are generally found in large groups, forming a microhabitat for a number of smaller animals, such as small crabs, that use the tubes for shelter and feed on organic debris from the worms. Vent clams and mussels are often found in association with tube worms.

Cold seeps

Cold seeps occur around all types of continental margin, in areas where methane and sulfide-rich fluids seep through the deep ocean floor. Although similar in many ways to hydrothermal vents, they are distinct—cold seeps release their cargo of chemicals at the same temperature as the surrounding water. Most cold seeps have been discovered when submersibles have come across pockmarks, or mud volcanoes, with communities of animals like those around hydrothermal vents. This similarity is no coincidence. Similar to the animals around hydrothermal vents, cold seep animals are dependent on symbiotic bacteria for their survival—the bacteria converts the methane and hydrogen sulfide into food for the seep community.

↓ **Tube worms at cold seeps** are smaller relatives of the giant tube worms found at hydrothermal vents. The tube worms form dense thickets that are home to mussels, small worms and crabs.

→ **Hydrogen sulfide** in cold seep water is absorbed by the blood-filled plumes on the top of these tube worms. These worms are often found in bushlike groups about 3 feet (1 m) in width and height.

→ **Large deep sea clams** at cold seeps depend on the food produced by symbiotic bacteria in their gills, which use the surrounding hydrogen sulfide. The hydrogen sulfide is poisonous to the clam and its blood contains specialized proteins that carry the hydrogen sulfide to the bacteria and prevent it affecting the hemoglobin in the blood. Those clams located farthest from a seep—where hydrogen sulfide is diluted—have proteins 10 times more efficient at picking up hydrogen sulfide than those from clams closer to cold seeps.

FORMATION OF COLD SEEPS

The methane, hydrogen sulfide and ammonia in seep water are produced by bacterial breakdown of organic matter in the sediments where there is no oxygen. In seeps below 656 feet (200 m), the surrounding water pressure and low temperature converts some of the methane into an icelike solid known as a clathrate. Unlike hydrothermal vents, cold seeps can be formed in several ways. The seeps in the Pacific trenches and off Florida were produced by the subduction of an ocean plate squeezing out the seep waters held in the sediments. In areas where there are substantial amounts of oil or freshwater deep within the sediments, these fluids tend to rise as they are less dense than seawater, and force out the seep water above them.

→ **These mussels are at the edge** of a brine pool. The water emerging from seeps can be so salty and dense that it does not mix with the surrounding water. Instead, it accumulates in depressions on the seabed, forming pools with mirrorlike surfaces. The water is so salty and devoid of oxygen, that animals submerged by rising brine will quickly die.

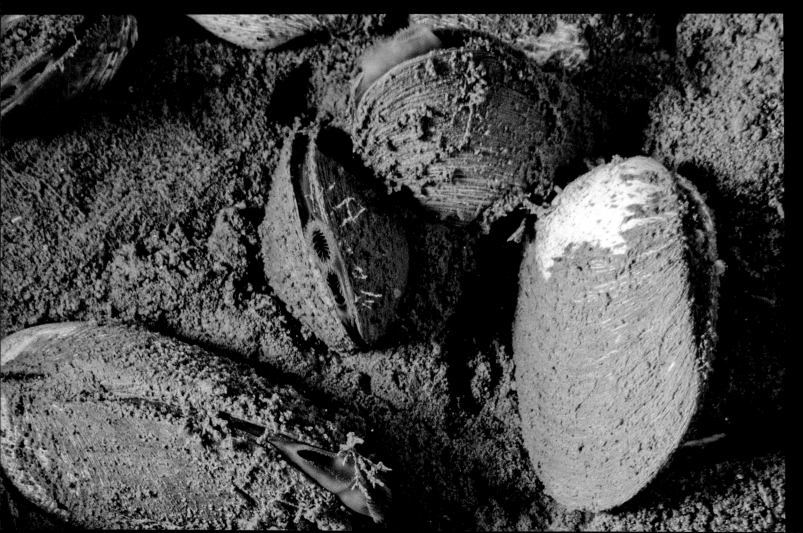

The fringes

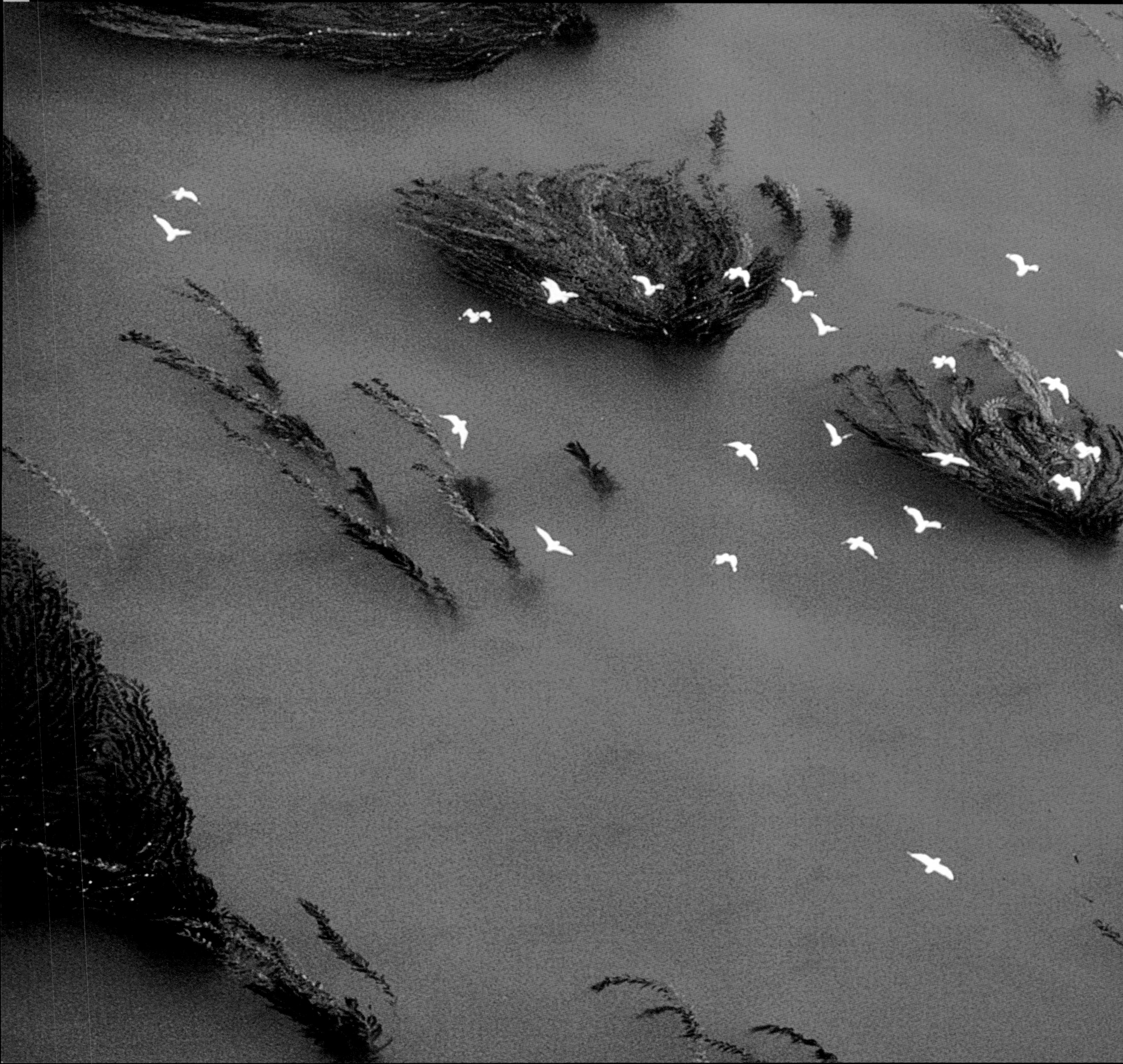

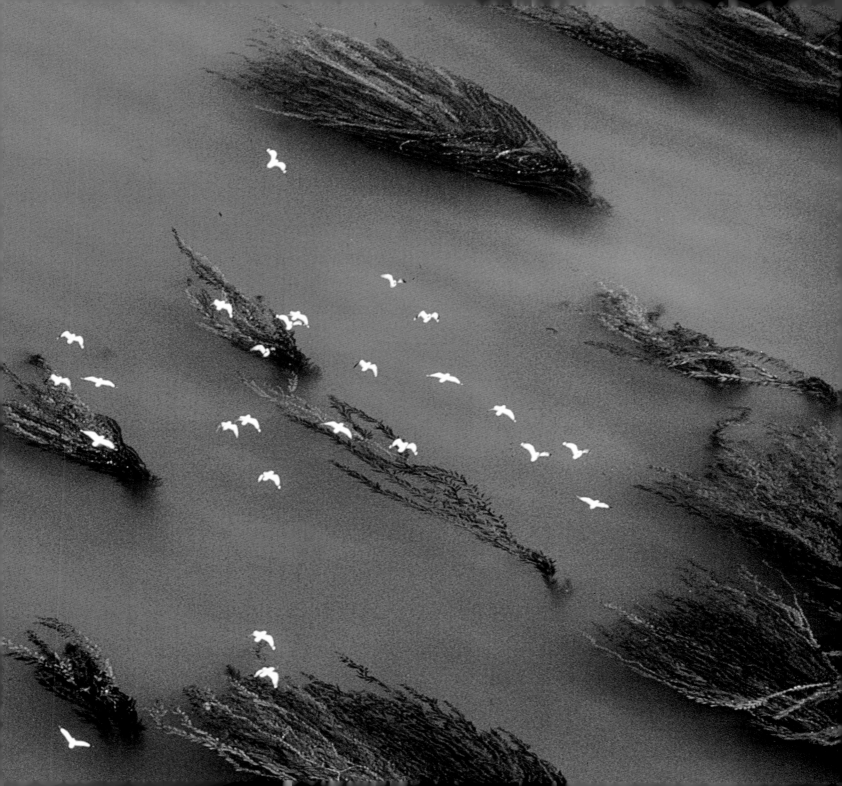

The fringes

Most marine life is found from the splash zone at the top of the shore down to the edge of the continental shelf. Among the varied environments of the fringes are coral reefs and mangroves—some of the most biologically diverse habitats on Earth.

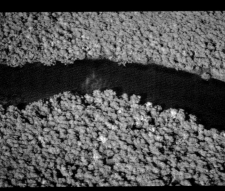

Coral reefs worldwide

Warm-water coral reefs rival tropical rainforests in terms of the range of plant and animal life associated with them. Fishes and invertebrates that live among the coral reefs are important not only as part of the marine ecosystem, but also as a vital source of food for human consumption. Their presence around tropical islands provides a buffer for coastal communities against the worst ravages of tropical storms. With the increase in mass air travel, coral reefs have become popular destinations for those seeking rest and relaxation on tropical shores or the thrill of diving into the world of the coral reef. Reef tourism provides a significant source of income and employment in some of the world's most impoverished countries. However, coral reefs are vulnerable to a number of threats from human activities.

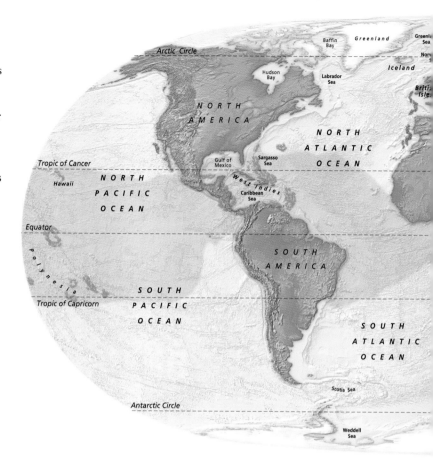

MOST THREATENED CORAL REEFS

Location of threatened coral reef hotspots

1 Philippines
2 Gulf of Guinea
3 Sunda Islands (Indonesia)
4 Southern Mascarene Islands (near Madagascar)
5 Eastern South Africa
6 Northern Indian Ocean
7 Southern Japan, Taiwan and southern China
8 Cape Verde Islands
9 Western Caribbean
10 Red Sea and Gulf of Aden

HUMAN IMPACT

Every year, large areas of reef are irreparably damaged by sewage and oil pollution, sedimentation and unsustainable fishing, as well as global warming which may be responsible for coral bleaching. Cold-water corals in temperate regions are also under threat from trawling.

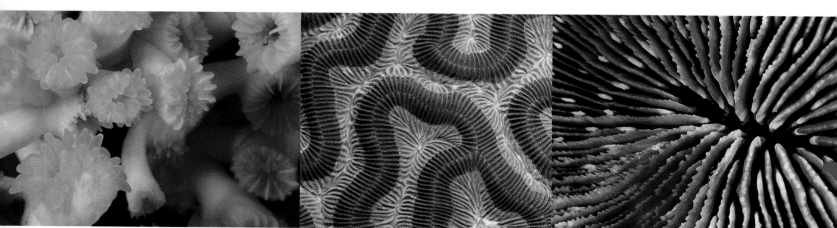

North Atlantic cold-water coral lives in deep water between 656 feet (200 m) and 6560 feet (2000 m).

Brain coral lives in shallow warm water. It gets its name from the massive rounded mounds it forms.

Mushroom coral is a warm-water species that forms huge underwater mushroom structures.

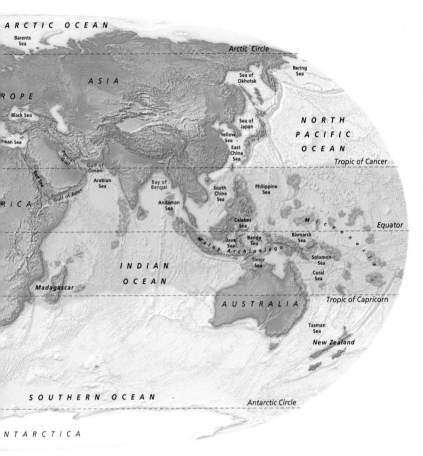

CORAL DISTRIBUTION KEY

Warm ocean

 Deep-water coral reefs

 Warm-water coral reefs

LARGEST CORAL REEF AREAS

Rank	Location	Coral reef area square miles (sq km)	Percentage of world total
1	Indonesia	31,700 (51,020)	17.95%
2	Australia	30,400 (48,960)	17.22%
3	Philippines	15,570 (25,060)	8.81%
4	France (French Overseas Departments)	8870 (14,280)	5.02%
5	Papua New Guinea	8600 (13,840)	4.87%
6	Fiji	6220 (10,020)	3.52%
7	Maldives	5540 (8920)	3.14%
8	Saudi Arabia	4140 (6660)	2.34%
9	Marshall Islands	3800 (6110)	2.15%
10	India	3600 (5790)	2.04%
11	Solomon Islands	3570 (5750)	2.02%
12	United Kingdom (British Territories)	3420 (5510)	1.94%

AGE OF CORAL REEFS

There is a relationship between the richness of life in a warm-water coral reef and its age. The oldest reefs have the greatest number of species. The reefs with the greatest number of fish species are those of the central Indo-Pacific region. More than 2000 fish species have been recorded on reefs in the Phillipines and at least 1500 species of fishes are found on the Great Barrier Reef, with nearly 800 from just one reef complex (One Tree Reef). The Atlantic is a much younger ocean basin, and this is reflected in the lower numbers of reef fish species.

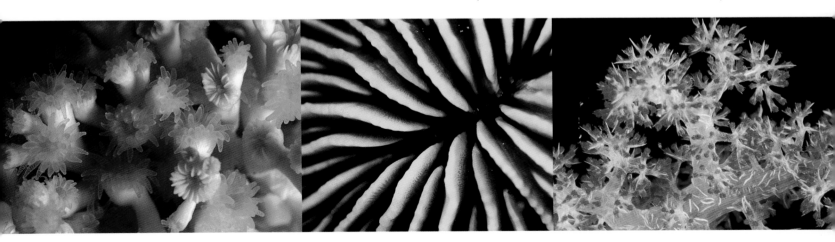

North Atlantic cold-water coral is found in white, yellow, orange and pink forms of the same species.

Mushroom coral has some of the largest individual polyps; they can reach 10 inches (25 cm) in diameter.

Coral polyps build tough structures around themselves for protection and support.

Coral reefs

Coral reefs have been growing in the world's oceans for more than 450 million years. They form massive structures that are not only fascinating biological communities, but also the largest geological structures on the planet that are formed by living organisms. With the exception of one or two cold-water species, the reef-building, or hermatypic, corals are extremely temperature sensitive. They can grow only where the water temperature does not fall below 70°F (21°C) or rise significantly higher. As a consequence, coral reefs are restricted to tropical waters (approximately 30°S to 32°N) and are not found on the westward-facing coasts of landmasses. These are areas of upwelling that bring cold, deep water to the surface, lowering the water temperature. Reefs are classified into three main types: fringing reefs, which form in direct contact with the margins of the land, usually on the downwind (low rainfall) margins of tropical islands; barrier reefs, which are separated from the shore by a lagoon or channel and are found around islands or parallel to continental coastlines; and atolls, which are ring-shaped coral islands circling a shallow lagoon.

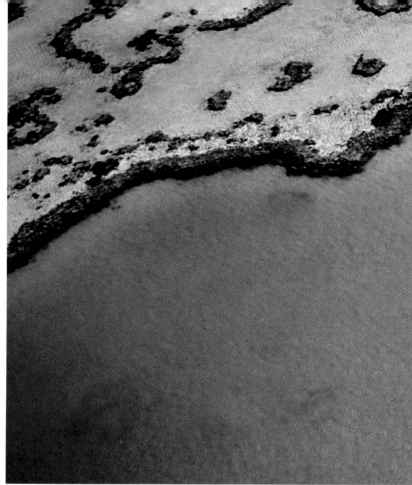

↓ **The aptly named Blue Hole is found in Belize, Central America.** It is a unique reef that cannot be placed into one of the three main types of formation. It was formed at the end of the ice age, some 10,000 years ago, when sea levels rose—this inundated a dry cave system and collapsed its roof. Coral colonies established themselves around the now-submerged hole and kept pace with rising sea levels.

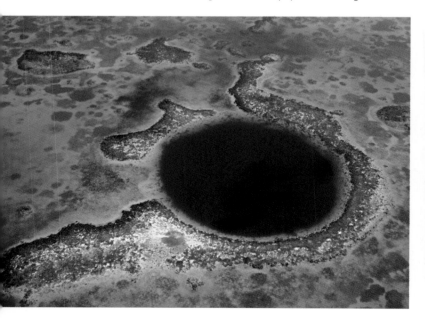

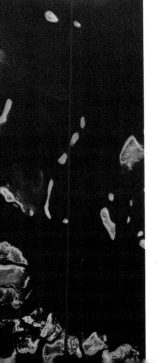

↑ **The biggest of all coral reef** structures, and the largest biological entity on the planet, is the Great Barrier Reef. It stretches for some 1560 miles (2500 km) along the northeastern coast of Queensland, Australia, and is an astounding 95 miles (152 km) at its widest point. It is not a single, continuous structure, but comprises thousands of interlinked segments that are oldest and thickest at the northern end.

→ **In both solitary and reef-building** hard coral species, the individual animals secrete a limestone skeleton that partially encloses the living polyp. The tentacles emerge through the top on the hard casing that is covered by a thin layer of soft tissue. The soft tissue layer is shared with surrounding polyps in reef-building species.

← **Photographing the Great Barrier Reef** from the space shuttle reveals its fragmented nature. It is a vast collection of single reefs, rather than a continuous stretch. The seaward face of the reef is at the base of the photograph.

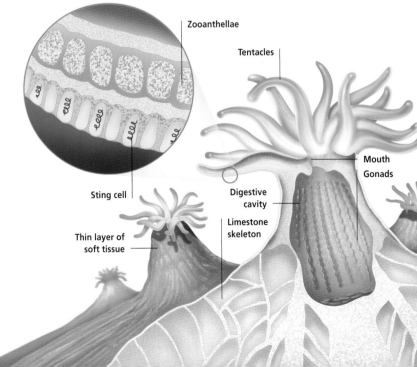

Zooanthellae

Tentacles

Mouth
Gonads

Sting cell

Digestive
cavity

Thin layer of
soft tissue

Limestone
skeleton

Coral communities

Tropical coral species tend to grow in areas of strong wave action. Such areas are rich in dissolved nutrients and suspended food, as well as rapidly dispersing waste matter produced by a large concentration of living organisms. Wave action breaks down dead coral to fine sand and opens up areas to new growth. There is always a balance between growth and destruction on a coral reef. Once established, it provides a habitat for an abundant and diverse group of plants and animals that live on, burrow into, and even eat, the hard coral. There is always strong competition between these reef residents for food, space and mates, and they must also protect themselves against predators. All these factors produce a community of animals characterized by bright colors, spines, extreme body shapes, camouflage, venoms and toxins.

CORAL BLEACHING

One of the most widespread, but poorly understood, problems that affect coral reefs is the phenomenon of coral bleaching. Affected corals expel the zooanthellae (symbiotic algae) that give them much of their color, resulting in white patches over the colony. Bleached colonies never lose all their zooanthellae. In some cases, there is a degree of recovery where new species of zooanthellae become viable within the coral polyps. However, it has been found that bleached colonies do not continue to grow and are more susceptible to being broken apart by wave action.

↓ **Loggerhead turtles have powerful horny beaks** well suited to cropping down the algal "turf" that covers most reef flats. Loggerheads, and other sea turtles associated with reefs, also feed on encrusting organisms such as sponges.

←← **The brittlestar feeds** on food particles and detritus trapped on the surface of brain coral. There is evidence to suggest that many species living on the coral are able to absorb the mucus and other organic matter that is constantly shed by the coral polyps. The brittlestar may be supplementing its solid food with dissolved nutrients.

← **These branching corals in the waters** around Fiji are often grazed by parrotfishes that have powerful beaklike teeth to bite off algal fronds or pieces of dead coral, which is then crushed to extract algal cells from the coral matrix, passing out as fine sand. In many reefs, parrotfishes make a substantial contribution to the coral sand around the reef.

↑ **This featherstar is attached** to a fire coral. These corals get their name from the powerful sting cell they use to ward off predators.

← **Not all large structures** on a tropical reef are corals. The sea fan is a relative of the corals but its body comprises tough proteins reinforced with chalky spicules. The individual polyps of the sea fan link to form a colony.

Atolls

An atoll is a ring-shaped island composed of a coral reef encircling a shallow lagoon devoid of anything jutting above the surface of the water. Some 300 atolls have been charted throughout the world, the majority of which are in the Pacific. They are usually found in groups, although some are isolated. They are found in all depths of water, from the continental shelf down to the deep open ocean. Most are about ½ mile (800 m) in diameter but the largest atoll, Kwajaleei, in the Marshall Islands, measures an impressive 176 miles (280 km) in circumference and encloses a lagoon with an area of 1100 square miles (2850 km²). All atolls are affected by prevailing wind and waves. On Pacific atolls, the windward side is colonized by encrusting algae that can endure the pounding of the waves. Depending on their size, atolls can support a wide variety of life. Initially, though, colonization begins with the establishment of plants that stabilize the coral sand.

→ **Bora Bora is a fringing reef**, rather than an atoll. There is still a remnant of the volcanic center in the lagoon. However, as the central remnant subsides, the fringing coral reef will continue to grow, eventually forming an atoll.

↓ **This group of small atolls** in the Maldives shows the typical circular shape and fully enclosed shallow lagoon of an atoll in the mid-phase of its development. Corals cannot grow within small atoll lagoons because the conditions are too extreme.

→ **This satellite image of the Maldives** shows a range of atolls at different stages of development. The small white dots are seamounts with fringing reefs that will ultimately become atolls.

⇒ **Kayangel Atoll, Palau, Micronesia,** is an example of a large atoll in the final phase of development. The original circular reef has become fragmented as the seafloor below has subsided and moved gradually toward a subduction zone.

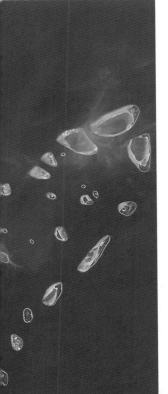

EMERGING VOLCANO

FRINGING REEF

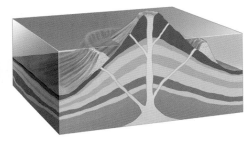

VOLCANO SUBSIDES

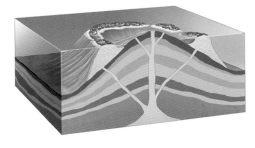

FORMATION OF AN ATOLL

Volcanoes emerge from the sea to form islands in tectonically active areas. Fringing reefs form on the sides of the volcanic island. The volcano will slowly sink down if it is on a subducting plate being drawn under another plate. Where the coral can grow upward, at the same rate as the sinking of the volcano, the volcano will eventually disappear, leaving the central lagoon of an atoll.

Shelf seas

From the earliest times to the present day, the shelf seas have been the focus of the majority of human maritime activities. Shelf seas are the shallow-water zones around continental margins and are the oceans' major source of biological and mineral riches. Indeed, some 90 percent of all commercially exploited fish species consumed in Europe and North America, come from shelf-sea fishing grounds. They extend from the low water mark to the edge of the shelf break. The average width of the continental shelf is 43 miles (70 km) but it can be much wider. For example, the continental shelf off the northern coast of Siberia extends 559 miles (900 km). There are two types of continental margin, whose size and character have a marked effect on the nature of the overlying shelf seas. The Atlantic type of continental shelf is broad, stable and gently sloping. The Pacific type is much narrower and is often close to areas of seismic activity. Despite the names, both types of shelf are found in other oceans.

↓ **Snappers are found** in large numbers in shallow waters throughout the tropics. In these parts, many inshore fishing communities depend on them.

Trevally are a group of shallow-water fish species that have become increasingly important in tropical and subtropical shelf seas for commercial and recreational fishing. Traditionally, they have been caught by rod and line, but, in recent years, inshore trawling has become more common. This great increase in catches poses a threat to some stocks, as trevally are slow-growing fishes.

Pacific herring were once abundant over the continental shelf around the Pacific Ocean. Today, commercial fishing, including egg collection, in Japanese coastal waters has drastically reduced stocks in some areas.

Coastal waters of shelf seas are important because of their value in attracting tourism.

SHELF SEA ELEMENTS

The transition from the edge of the sea to the ocean depths follows the same pattern around all continental margins, although relative widths and depths vary greatly. The shelf zones are a continuation of the continental landmasses. For example, on mountainous coastlines, shelf zones are narrow, rough and steep, but smooth and gently sloping where plains meet the sea. Continental shelves are covered with sediments from rivers or blown from the land. The shelf edge is marked by an abrupt increase in gradient called the shelf break; in some places, steep-walled V-shaped submarine canyons cut deeply into both the shelf and the slope below. The continental slope drops steeply to the seafloor, marking the outer boundary of the continental landmass. At the bottom of the slope, where the landmass meets the ocean crust, there is often a gently sloping zone between the slope and the abyssal plain, called the continental rise.

SHELF SEA CROSS-SECTION

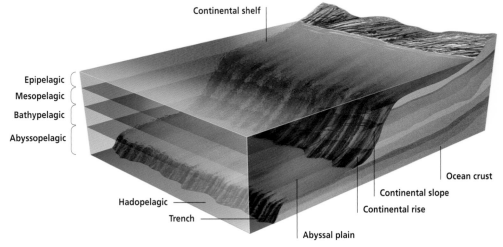

Continental shelf

Epipelagic
Mesopelagic
Bathypelagic
Abyssopelagic

Hadopelagic
Trench
Abyssal plain
Continental rise
Continental slope
Ocean crust

Estuaries

Estuaries are sites where the sea not only meets the land but also mixes with freshwater, giving rise to environments that have a highly specialized set of plants and animals. They have also become sites of human settlement and exploitation—many of the world's great cities, such as London, Tokyo and New York, are sited on estuaries.

Estuaries are formed in four ways (see illustrations, *right*), and can also be classified according to levels and interaction between fresh and saltwater. Many owe their existence to the rises in sea level that followed the end of the last ice age 10,000 years ago, drowning coastal river valleys. Estuaries are also formed where coastal currents form barrier islands and sandbars offshore with a brackish lagoon behind. Tectonic estuaries, such as San Francisco Bay (*right*), are created when local faulting and subsidence cause the inundation of land below sea level by fresh and saltwater. Fjords are a form of estuary that are produced by glacial action.

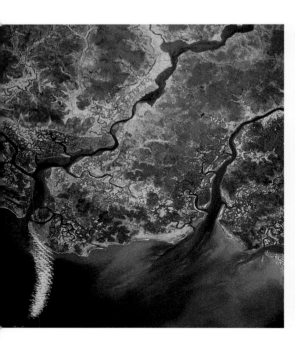

DROWNED RIVER VALLEY ESTUARY

These form when existing river valleys are flooded by seawater, and sedimentation rates do not keep up with rising sea levels. They retain a typical V-shaped cross-section. They often have extensive mudflats.

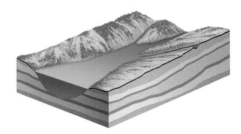

FJORD

Fjords formed 18,000 years ago, at the beginning of the last ice age. Fjords are deep and sheer-sided, as the flow of ice down pre-existing river valleys deepened and widened them. In most fjords, the river flow is small, with little sediment.

BAR-BUILT ESTUARY

Beginning as drowned river valleys, here the sediment keeps pace with sea levels. These estuaries occur where there is a large river flow and sediment burden, which increases in periods of flood. This sediment results in the creation of a bar and lagoon.

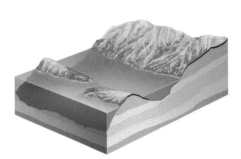

TECTONIC ESTUARY

These are the rarest of the main estuarine types. They are created where tectonic faults cause land at the coast to drop below the surrounding area. When the sea breaks through, and with freshwater input, an estuary is formed.

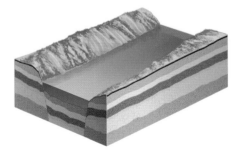

← **The estuary in San Francisco Bay** owes its origins to the seismic activities associated with the nearby San Andreas Fault.

← **These two estuaries,** on the west coast of Africa, are examples of the best-known form of estuary—the coastal plain, or drowned river valley, estuary.

Estuarine life

Only those plants and animals that can cope with a perpetual variability in temperature and salinity can successfully exploit estuarine habitats. All organisms have to maintain a balance of salts and water in their cells, but in an estuary, the task of maintaining this balance is exacerbated by the rapid changes of salinity. As a result, estuarine communities have very few species in comparison to either the adjacent freshwater or marine coastal habitats. Most estuarine species have marine origins. Those species that can tolerate the extreme estuarine conditions benefit from reduced competition for space and food, and they have fewer predators to contend with. Also, not all estuarine species are permanent residents. Many species of coastal fishes, for example, use estuaries as nursery grounds, ensuring that vulnerable young can grow in relative safety, away from most predators.

→ **The abundance of invertebrates in estuaries** supports large numbers of fishes, particularly juveniles, which attract fish-eating inshore birds.

↓ **Juvenile eels,** born at sea, pass through estuaries on their way upriver to adult life in freshwater.

OSMOREGULATION IN FRESHWATER

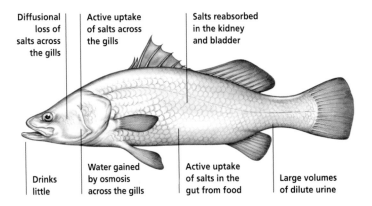

Diffusional loss of salts across the gills

Active uptake of salts across the gills

Salts reabsorbed in the kidney and bladder

Drinks little

Water gained by osmosis across the gills

Active uptake of salts in the gut from food

Large volumes of dilute urine

OSMOREGULATION IN SEAWATER

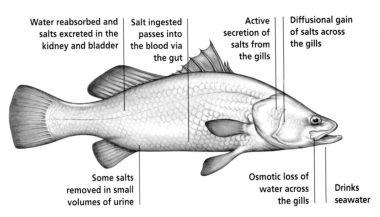

Water reabsorbed and salts excreted in the kidney and bladder

Salt ingested passes into the blood via the gut

Active secretion of salts from the gills

Diffusional gain of salts across the gills

Some salts removed in small volumes of urine

Osmotic loss of water across the gills

Drinks seawater

OSMOREGULATION

Osmoregulators actively control the composition of their body fluids, keeping it within a narrow range. A fish in freshwater gains water by osmosis, and when it feeds. It will lose important salts that diffuse out across the gills. To counteract these effects, fishes actively reabsorb sodium and potassium from their urine and offset losses at the gills by uptake of salts from the water. They remove the excess water by producing large amounts of dilute urine. However, fishes in saltwater will tend to lose water from their bodies and gain excess salts that need to be removed. The salts are actively pumped out by the gills and kidney. The kidney also reabsorbs most of the water from the urine so that the fish produces small volumes of concentrated urine.

→ **Saltwater, or estuarine, crocodiles** are the largest living reptiles. They exploit estuaries to the full—they not only feed on aquatic life, but also penetrate inland.

Fjords

Fjords are estuaries formed by the action of glaciers and occur only in high latitudes. In the northern hemisphere, they are found in Norway, Alaska and eastern and western Canada. In the southern hemisphere, they are seen in Chile and the South Island of New Zealand. During the last ice age, thick glaciers slid slowly down the mountains, carving deep, U-shaped channels into the river valleys and excavating the valley floor usually to well below sea level. The glaciers melted at the point of contact with the sea and further erosion was prevented. Large amounts of material that the glaciers picked up on their journey to the sea were deposited at this point, forming rocky barriers called sills. When glaciers retreat, the sea will flood into a deep valley that is partially blocked at its seaward end by a sill. As a result, dense seawater is often trapped in the bottom of fjords, while less dense freshwater from streams forms the upper layers. Water on the fjord floor is deoxygenated and cold, and little life exists there. Where stagnation is extreme, hydrogen sulfide is produced by bacteria that survive only in anoxic, or oxygen-poor, conditions. These conditions are similar to parts of the deep sea, and provide insight into how these less-accessible ecosystems work.

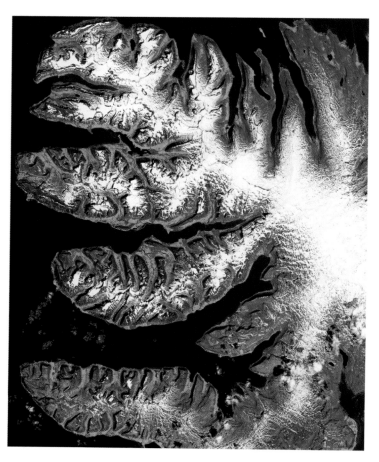

→ **Coastlines at high latitudes** are often dominated by fjords; settlement in these parts is restricted. This satellite image of the West Fjords peninsulas of Iceland provides a clear example.

↓ **Milford Sound in New Zealand** is one of the best-known fjords in the southern hemisphere. The conditions on this coast enable what are normally deep-water species to flourish—this is known as deep-water emergence.

← **The extreme depth** of this striking Norwegian fjord provides some indication of the vast size of the glacier that carved it out. Most of the shoreline of the fjord is sheer rock face. The only usable flat land lies where there is a small delta created by the river that now flows into the fjord's head.

↓ **Bridal Veil Falls** are located near the head of Geirangerfjorden. This fjord is considered by many to be the most beautiful in Norway. The falls were created by a glacier cutting across a pre-existing river valley which now hangs high up on the side of the fjord. Half the classic U-shape of a glaciated valley can be clearly seen.

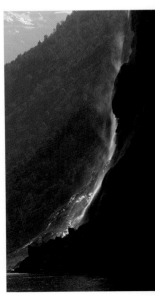

Mangroves

Mangroves are a feature of many tropical and subtropical estuaries and can form extensive fringing coastal forests. There is a wide variety of mangrove trees, ranging from small shrubs to large trees. They are adapted to living in soft, waterlogged, often anoxic (oxygen-poor) mud, as well as a daily inundation of saltwater. Mangroves are a habitat type rather than a strictly defined set of species, however, they all share one overriding characteristic—mangrove trees and shrubs cannot survive temperatures that fall more than 18°F (10°C) below their ideal of 66°F (19°C). As a consequence, they are restricted to the climates of tropical and subtropical estuaries where they can be found on both the westward- and eastward-facing coasts. Mangroves form spatially complex three-dimensional habitats with an extremely varied set of animal populations, each exploiting different aspects of mangrove ecology.

LIFE IN THE MANGROVES

As mangroves grow in extent, there is a gradual increase in the stability of the sediments on the landward side. In the oldest part of the mangrove, the mud is solid enough for burrowing species such as fiddler crabs and mud lobsters to dig permanent holes. On the firmest mud, animals such as crocodiles are able to move onto land, and the larger land animals can also feed on the margins. Moving toward the seaward fringe of the mangrove, the mud becomes softer and only animals that can cope with such conditions can flourish here. For example, the tropical cockle has ribs and spines that stop it sinking into the soft sediment, and the window oyster has a large, very thin shell that spreads its weight like a snowshoe across the mud.

↓ **The presence of mangroves in tropical** estuaries and adjacent coastal waters is important in the stabilization of low-lying coastal zones. The binding effects of the mangrove roots prevent coastal erosion by storms and currents, and help to reduce the risk of flash floods during the monsoon.

↘ **Cormorants in a Florida mangrove** roost in the tops of buttress roots and emerging young trees when the tide comes in. This helps to isolate them from land predators. At low tide, wading birds, reptiles and some mammals move into the mangroves where the mud is sufficiently firm.

← **Buttress roots and pneumatophores**—erect roots exposed to air—make mangroves the ideal nursery grounds for many fish species. Small fishes living in mangroves have abundant food, while agile prey species can escape their pursuers by hiding among the roots in spaces too small for predators to follow.

↓ **Aerial roots that carry** air down to the bottom roots of mangrove trees form vertical microhabitats with distinctive zoned communities of animals. Position on the root is determined by an animal's tolerance of tidal exposure. Here, a mangrove crab feeds on small barnacles.

Kelp forests

Kelps are a form of algae. As with mangroves and reef corals, their distribution is determined by surface water temperature. Kelp forests grow only in cold, nutrient-rich water. They are found mainly in the higher, temperate latitudes on the western sides of continents, where upwelling and current systems convey this type of water. Most kelp forests tend to be situated some distance offshore and can be as deep as 130 feet (40 m). Both these features of their location help the kelp to avoid excessive shearing forces caused by wave action.

Kelp forests are similar to their terrestrial counterparts in that they have a complex three-dimensional structure. Beneath a wide canopy of kelp lies a dense understory of smaller seaweeds that have adapted to low light levels. Many different animals shelter between the fronds or firmly attach themselves to the kelp itself.

→ **The resemblance to a forest on land** is clearly evident in this giant kelp bed. Kelps begin life as microscopic spores produced in special tissue on the mature adult plants. The giant kelp can grow to more than 165 feet (50 m).

↘ **The Californian big-bellied seahorse** uses its tail to anchor itself in the fronds of a giant kelp. The fronds give protection against excessive wave action and, in combination with the seahorse's coloration, help camouflage it from predators.

↓ **Banded chink shell snails** come together in abundance to graze on a small kelp species called oar weed. The snails will gradually rasp away the fronds.

← **The Californian bat ray** feeds on the mollusks, crustaceans and worms that live on the seabed around the holdfasts of the giant kelps. Bat rays dig up their food with their snouts. They have powerful, crushing jaws that they can use to bite off overhanging ledges to get at food.

↓ **This nudibranch sea slug** works its way over the flexible stem, or stipe, of a small kelp species called an oar weed. The sea slug is more delicate in its feeding than some of its mollusk relatives. It sucks off the mucus layer produced by the weed, rather than rasping away living plant tissues.

WEB OF LIFE

Sea otters were hunted to near-extinction for their fur. Although protective measures were introduced in 1911, population recovery has been slow. The massive reduction in sea otters impacted on their habitat, the giant kelp forests. Red and purple sea urchins, a staple food of the sea otter, are the only significant grazers of giant kelp, and normally feed on pieces that have broken off. However, when sea urchin populations explode, as they do on occasion, the echinoderms become so numerous that they resort to eating the living plants. Such plagues will go largely unchecked until the urchin-loving sea otter increases in number.

→ **Sea otters collect clams and urchins** from the seabed. They surface and then wrap themselves in thick kelp fronds as protection from predators, and to stop themselves from drifting away. In order to eat their prey, sea otters often lay a flat stone on their chest, using it as an anvil on which to smash open the clams and urchins.

Sea grasses

Sea grasses are flowering plants with roots, leaves and stems that distribute their pollen in seawater currents rather than by air or via insects. They are found throughout the coastal margins of temperate and tropical zones and are probably some of the most productive habitats on Earth, producing about 3½ ounces of new plant tissue per square foot (1 kg/m²) each year. There are about 50 species of sea grasses and all are found on mud, or muddy sands, from the low watermark downward, often forming luxuriant beds or meadows. The maximum depth at which they can grow is determined by the clarity of the overlying water. Both sea-grass beds as a whole and individual plants within them are important marine habitats.

In temperate zones, the principal sea-grass species are the eel grasses. They are sometimes found exposed on the lower shore but can grow as deep as 100 feet (30 m). The turtle and manatee grasses are the principal sea grasses in tropical and subtropical areas. They are found from the edge of the shore down to 33 feet (10 m), depending on conditions.

→ **Green surf-grass plants are the biggest** and most vividly colored of the sea-grass species. They have extensive and extremely tenacious root systems that anchor long, flexible but tough stems into the substrate. The tightly packed stems dissipate wave energy, enabling the plants to survive in the turbulent intertidal zones of temperate east Asia and northwest America.

↘ **This small crab in a temperate sea-grass bed** off the coast of British Columbia, Canada, is sheltered by the sea grass from excessive wave action and benefits from the food the bed provides. The crab feeds on grazers such as snails that can live in abundance on the leaves of the sea grass.

↓ **A Pacific giant octopus forages in a sea-grass bed** off the west coast of Canada. It is one of the top predators of the sea-grass community. Its prey includes a variety of animals such as sea urchins, crabs, shrimp and small fishes that find food and shelter among the sea grass. The octopus can hide behind grass stems and, with its long, flexible arms, reach round to grab unsuspecting prey.

LIFE IN THE SEA GRASSES

Sea grasses are important not only as primary producers in their own right, making a major contribution to local food resources for grazing animals, but also as extremely valuable microhabitats. The leaves of sea grasses are covered with a film of single-celled algae, particularly diatoms, that increase the overall primary production in the sea-grass bed. The leaves also carry blue–green algae that release nitrogen-rich nutrients that are essential for healthy plant growth. Sea grasses are grazed by animals ranging from sea turtles, manatees and birds, to sea urchins and starfishes. Dead sea-grass material becomes the food of many animals, such as worms and snails, that live in the mud around the plants.

→ **The bat star, seen here in a surf-grass meadow** off the Pacific coast of California, is exclusively a herbivore. It moves slowly over the sea-grass "turf," grazing off the film of diatoms, algae and bacteria without harm to the sea grass itself. Its adhesive tube feet are ideal for climbing on the smooth sea grass.

↓ **This sea slug is able to exploit** an abundant food supply in the form of the film of micro-organisms that cover the sea grass. Similarly, small fishes can find ample food and shelter in the dense sward of leaves and stems.

Sand and shingle shores

Even the most exposed shores contain pockets of small particles. In temperate latitudes, these are silica and other rock minerals, but beach materials can also be biological in origin, such as the dazzling white coral sand beaches in the tropics, or even black volcanic sands. Whatever their composition, these sand and shingle, or particulate, shores are constantly kept in motion by the ceaseless breaking of waves. This movement gradually rounds off particles and also sorts them, so there is a gradient of grain sizes across a beach that will affect properties such as water retention.

At first glance, sand and shingle shores appear barren and uniform in comparison to the richness and diversity of rocky shores. However, a closer examination of the organisms found on, or in, such shores reveals clearly identifiable patterns of zonation that can be related to physical conditions. These change as one moves across the shore. Particulate shores appear devoid of life mainly because of one major difference from rocky shores—they are inherently unstable and unsuitable for most species that require the stability of a solid rock surface. Most sandy shores are, therefore, mainly populated by organisms that are either big enough to stabilize the surface in their immediate vicinity, or live buried in the more stable lower layers. Few organisms can cope with such demands, but those that do are often found in abundance.

SLOPE OF BEACH	
Beach particles	Beach slope
Very fine sand	1°
Fine sand	3°
Medium sand	5°
Coarse sand	7°
Very coarse sand	9°
Granules	11°
Pebbles	17°
Cobbles	24°

BEACH SHAPE

On most beaches, the size of the sand or shingle particles directly relates to the slope of the beach (see table, *left*). In general, the finer the particles, the flatter the beach.

Regardless of particle size, beaches generally have the same cross-section, as shown below. The backshore is a relatively quiet area where particles move little and are often stabilized by plants. The berm marks the limit of sediment deposited by wave action, and its crest is often the highest point on the beach. Wave action at the base of the berm creates a scarp. This marks the top of the foreshore and the start of the expanse of the low tide terrace, which runs down to the low tide mark. Below low water, longshore currents and wave action create a longshore trough, sometimes accompanied by irregular longshore bars.

TYPICAL SAND OR SHINGLE BEACH

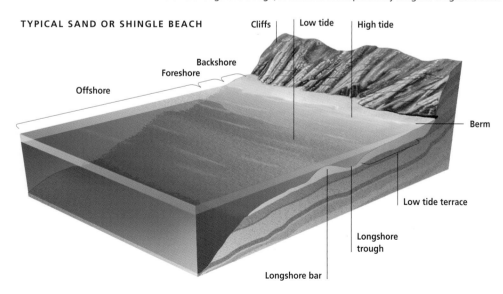

Cliffs | Low tide | High tide

Backshore

Foreshore

Offshore

Berm

Low tide terrace

Longshore trough

Longshore bar

⤒ **Coral sand beaches** owe their startling white appearance to fragments of the chalky skeletons of countless reef-building coral polyps. As coral reef colonies die, even moderate wave action will quickly break them down into fine particles and carry them to the shore.

↑ **Shingle shores are difficult for plants** and animals to colonize. The large and irregularly shaped pebbles create large air spaces, and make the beach free-draining at low water, trapping little potential food. Also, the pebbles' size and shape allow wave action to easily move them, so that they grind against each other.

← **Coastal sand dune systems are formed** where prevailing onshore winds winnow off sand from the drier parts of the beach. Dunes reach a maximum size that is determined by a combination of factors. However, grain size, sand supply and wind strength are by far the most signicant determinants.

Rocky shores

Rocky shores present some of the most dramatic interactions between land and sea. The spectacle of waves crashing over a rocky shore reminds us of the power of the sea. When the waves recede, the abundance and variety of lifeforms displayed have fascinated countless generations. One of the most interesting features of rocky shores is the consistency of certain characteristics around the world. In particular, it has long been known that the distribution patterns of rocky shore plants and animals from different parts of the world are remarkably similar. Even the most cursory examination will reveal that, although individual species differ, there are distinct bands, or zones, at particular vertical heights above low water that are always similar in composition.

The physical geography of rocky shore coastlines is determined by a number of factors. Among these is the nature of the rock type and its susceptibility to mechanical and chemical weathering and wave action. For example, smooth igneous granites are much more resistant to erosion than faulted, softer sedimentary or metamorphic rocks.

STRAIGHTENING THE COASTLINE

When a coast is exposed to the powerful erosive forces of the sea, these forces will initially intensify the irregularities along the coastline. Rocks are rarely uniform in composition over any appreciable horizontal distance, and wave action will pick out less-resistant areas. This gives rise to geologically ephemeral structures such as arches, stacks and retreating cliffs that leave wave-cut platforms. However, as these structures disappear, the coastline starts to smooth out. The concentration of wave energy on headlands increases erosion on this part of the coast, but the dissipation of wave energy, as wave fronts spread out, creates areas of sediment deposition. Eventually, the sediments may spread sideways and protect the base of the headland, slowing or stopping further erosion and remaining in dynamic equilibrium, neither losing nor gaining more sediments. The timescale of this straightening process is directly proportional to the intensity of the wave action.

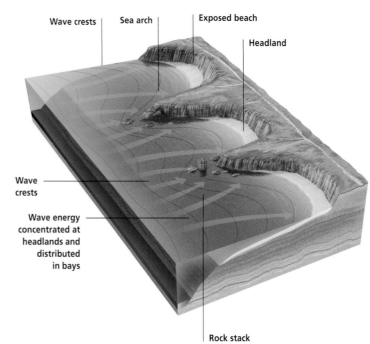

Wave crests   Sea arch   Exposed beach

Headland

Wave crests

Wave energy concentrated at headlands and distributed in bays

Rock stack

← **The Twelve Apostles on the coast** of Victoria, Australia, are geologically short-lived features. They first appeared as headlands, then became sea arches and are now isolated sea stacks. The weaker rock around them has been gradually eroded by the constant pressure of wave action. They will eventually succumb completely to the power of the sea, and end up as sand.

←← **The geology of rocks exposed** to wave action has a marked effect on the shore ecology. Here, horizontal bedding planes in the sandstone have been picked out by waves and eroded into long crevices and extensive tide pools.

Life in the strandline

All beaches reveal where the last high tide deposited its load of dead seaweed, animal remains and other marine detritus. This is called the strandline and, if the material has been on shore for some days, it is not the most appealing of habitats. However, the strandline is one of the main sites of biological activity on sand or shingle shores. Corpses and decaying weed provide an abundance of food and moisture, both of which are in short supply in the upper parts of such shores; they are, therefore, a natural focus for animal life on the surface of the shore.

In temperate zones, turning over a pile of weed on the strandline is likely to produce an eruption of beach hoppers and sand fleas. These are small crustaceans that live in burrows beneath the strandline during the day, protected from the Sun and the risk of drying out. They emerge at night to feed on the organic material. Living in a burrow means that these crustaceans have no visual cues to help them distinguish between night and day. Instead, they have developed sophisticated internal clocks that ensure that they only emerge during the hours of darkness.

↗ **Marram (or beach) grass is found from the highest** strandline points on sandy beaches, extending back into dry sand or dune areas. The grass spreads its roots just under the surface of the sand and helps to keep the sand in place.

↑ **The sand hopper, or sand flea,** is actually a crustacean that abounds on temperate shore strandlines. It is nocturnal and a strong jumper—it is able to leap many times its body length into the air, mainly as an escape response.

→ **On tropical and subtropical shores,** ghost and fiddler crabs sift the sand, picking out edible material and rolling the rejected particles into small pellets.

⇉ **Tropical ghost crabs scurry** along the strandline in search of food deposited by the sea. Older, larger crabs live high up the shore and smaller, younger crabs avoid becoming a meal by living closer to the water's edge.

↓ **Strandline material** on an Australian beach shows some of the various important components that support abundant life. Decaying corpses and dead animal tissue provide a rich but short-term supply of food and moisture. The decay of plant remains is much slower, but often produces patches of moisture-retaining soil. These are not only refuges for strandline animals, but also help the establishment of terrestrial plants, such as tufts of marram grass.

Shore life

Many animals that are not strictly marine can be found feeding on rich pickings at the margins of the sea, either on shore plants and animals or on the detritus of the strandline. Bears are often seen looking for food on temperate and subpolar northern shores. Their wide-ranging diet includes carrion and invertebrates. A keen sense of smell guides them to large carcasses that have washed up, and their powerful front legs make it easy for them to dig up buried shellfishes, sea urchins and large worms. On tropical shores, many primates, such as the crab-eating macaques, live and feed primarily on the landward side of mangroves and beaches. This style of feeding, and the many similar features between human and primate bodies, have led a number of scientists to suggest that modern humans may have evolved as shore animals. Today, in many parts of the world, humans still collect shore plants and animals as food.

↓ **An Alaskan brown bear** uses its powerful paws to dig in the sandy shore for a favored meal of fast-burrowing razor clams. A red fox watches in anticipation of picking up any edible scraps that the bear leaves behind.

→ **Despite their name, the preferred food** of crab-eating macaques is the abundant fruit found in the drier parts of mangroves. They will also eat some larger invertebrates, such as the horseshoe crab, which can be found in large numbers in shallow waters.

In many parts of the world, shore plants and animals are collected for human consumption. The rich salt-marsh pastures of Normandy, France, provide the much-prized pré-salé lamb which has a unique salty flavor. These sheep are grazing on marsh samphire, close to Mont-Saint-Michel.

→ **Springbok graze on sparse dune** vegetation on the Skeleton Coast of Namibia. This is one of the driest parts of the world. Rainfall is unpredictable and vegetation must rely on the moisture-laden winds blowing from the sea for sustenance. The Benguela Current flowing along the coastline helps to keep the air cool.

Tide pools

Tide pools are refuges for many species unable to cope with exposure to air. However, although the plants and animals can remain covered with water as the tide recedes, life in a tide pool imposes its own severe demands. Once exposed, tide pools are in contact with air, which often has a significantly different temperature to that of the sea. The rate of warming or cooling of the tide pool is dependent on its size—the temperature of a large pool changes more slowly than that of a small one. Position on the shore greatly affects the variety of life that can inhabit a tide pool. Simply because of the difference in exposure times, pools sited high on the shore present a greater physiological challenge than those close to the low watermark. Other conditions also change. The oxygen content drops as it is used in respiration, while an increase in water temperature decreases the amount of oxygen that the water can hold. Carbon dioxide buildup makes the water acidic. Salinity in tide pools is subject to significant changes. Evaporation by heat increases the salinity; rain reduces it.

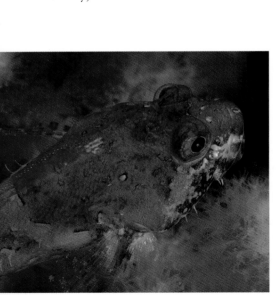

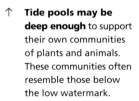

↑ **Tide pools may be deep enough** to support their own communities of plants and animals. These communities often resemble those below the low watermark.

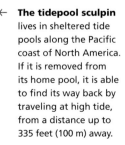

← **The tidepool sculpin** lives in sheltered tide pools along the Pacific coast of North America. If it is removed from its home pool, it is able to find its way back by traveling at high tide, from a distance up to 335 feet (100 m) away.

LIFE IN A TIDE POOL

Hermit crabs occur in tide pools everywhere. They live in the abandoned shells of mollusks

Seaweeds are marine algae that anchor themselves by means of a suckerlike structure

Barnacles open their carapace and extend their limbs to feed

Small octopuses are quite common but often elusive

Sculpin are common in tide pools and are also found in deep water

Mussels often grow in dense clumps or colonies

Blennies lack the swim bladder common to fishes in the open sea

Starfishes are carnivorous, mainly preying on mollusks

Sea anemones feed on smaller tide pool animals

Small crabs are often abundant in tide pools

FISH ADAPTATIONS

Tide-pool fishes rarely exceed 8 inches (20 cm) in length, and have slender bodies. Small, thin body forms are thought to reduce the effects of turbulence when a tide pool is inundated by the incoming tide, and allow fishes to shelter in little cracks and crevices. Fishes may remain on the bottom of the pool to avoid wave action and usually have thickened skin over the fin rays to cope with the abrasion from rough surfaces. Fishes that do not hide often have means of clinging to weed or bare rock. For example, tide-pool seahorse species hook their tails around weed clumps.

← **These urchins graze encrusting organisms.** The spaces cleared by the urchins are re-colonized by animal larvae or seaweed spores washed in on high tides. As a result, there is a constant change in the composition of the tide-pool community.

Life in mud

Muddy sediments are found in sheltered coastal waters, where fine particles brought down by rivers are precipitated out when freshwater meets seawater. Here, they settle into a fine layer on the seabed. As a habitat, muds have a number of complications. In their favor, they remain wet when the tide is out, they are easily burrowed into for shelter, and they contain large amounts of potential food. Muds in estuaries are also favored by animals, such as the polychaete worms, that use them to avoid the rapid changes of salinity in the overlying water. Water penetrates slowly into mud; salinity changes are, therefore, slow and greatly reduced in range. However, mud dwellers have to overcome considerable problems. First, in addition to inorganic particles, large amounts of organic matter are deposited at the same time, which will decay and use up oxygen. Secondly, oxygenated water cannot penetrate more than about 1 inch (2.5 cm) into muds. The combination of these factors means that only the top layers of the mud contain oxygen. Sediments below the top quickly become more anoxic with depth. The lack of oxygen allows the growth of bacteria, which produces hydrogen sulfide. This poisonous chemical reacts with iron salts, turns the mud black, and gives it an unpleasant rotten-egg smell. Animals are able to live in mud either by maintaining contact with the surface or by adapting to life with little or no oxygen. Some even make use of the hydrogen sulfide, in the same way as deep sea vent animals.

→ **Mudflats are sites of intense** biological activity. Their surfaces are covered by films of organisms that support surface grazers. The mud also carries organic matter that feeds many of the buried filter feeders.

↓ **This cast was created** by a lugworm that lives in U-shaped burrows in the sand. The worm eats sand, dimpling the surface above its head, and digests the organisms that live between the grains. Every few hours, it pushes out the unwanted sand to the surface.

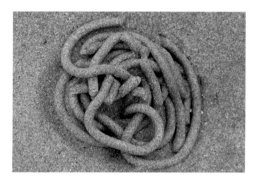

↑ **Moon snails are active hunters,** diving into the mud to find small mollusks and worms. Their shells are protected against abrasion when moving through the mud, by a cloak of soft tissue that is an extension of the front of the foot. Paradoxically, it is easier for the snail to regenerate soft tissue than hard shell.

← **Cerith snails are found** in vast numbers on mudflats. They are able to use their broad foot to spread their weight and glide over the mud surface. They feed on the films of single-celled algae and diatoms that live in the water left by the tide. The snails feed at low water, but burrow into mud when the tide turns.

↑ **Mudskippers get their name** from their habit of lashing their tails so that they move across the mud in a series of skips. They spend most of the time out of water, and breathe air by using the thin lining of their gill chambers as a simple lung.

MUD COMMUNITIES

At first glance, animals found in shallow-water marine mud appear to be a random assortment of species. However, this is far from true. Studies have shown that these animals form communities with an identifiable structure and they are classified according to the most abundant or dominant species. In this illustrated community (*below*), there are surface grazers (mud snails), predators that feed from the surface (moon snails), and buried worms, crustaceans and bivalve mollusks that all depend on food deposited on the surface, or in suspension immediately above it. The dog whelk is the only exception, feeding on buried bivalves.

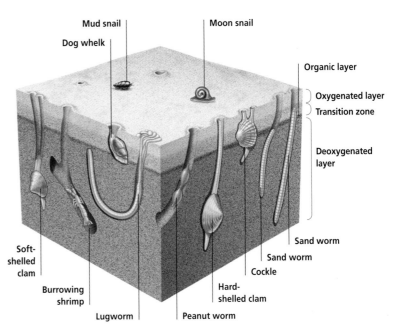

Mud snail

Moon snail

Dog whelk

Organic layer

Oxygenated layer

Transition zone

Deoxygenated layer

Sand worm

Sand worm

Cockle

Hard-shelled clam

Soft-shelled clam

Burrowing shrimp

Lugworm

Peanut worm

Life around ice

Polar food chains are essentially marine. Only a few top predators are land-based. However, there are differences between the poles. The Arctic is an ice island accessible to large land predators, whereas Antarctica is an isolated continent that no longer has any terrestrial animals.

There is a diverse and abundant community of animals beneath the winter ice of the Antarctic shelf. These animals are found on the seabed where the water is deep enough to avoid scouring by the ice. The richness of life is probably attributable to the relative constancy of the conditions. Antarctic water temperatures remain throughout the year within a very narrow range, usually between 32°F and 39°F (0°C and 4°C). Life proceeds without seasonal checks, albeit at a slow pace. Life under ice also favors the growth of glass sponges that contain symbiotic algae. Their long spicules are thought to act as fiber–optic conduits, carrying the limited amount of light to the algal cells. The extreme conditions under the ice mean that the communities that have developed there have more in common with creatures inhabiting the deep sea than other shelf areas.

→ **Chinstrap penguins live on the ice** around the edge of the Antarctic peninsula and nearby islands. They are also found on the larger icebergs that break off from the Antarctic ice. They live in large colonies, feeding on the abundant krill and fishes of the Southern Ocean. The adult penguins, in turn, are preyed upon by leopard seals, while their chicks and eggs provide food for sheathbills and brown skuas.

↘ **Crabeater seals below the sea ice** around Signy Island, Antarctica, must remain in areas where there are open holes for them to surface and breathe. Despite their name, they do not eat large crabs. Instead, they have highly modified teeth that enable them to strain out large zooplankton, mainly eating krill. In the water, the adults may be taken by killer whales.

↓ **Antarctic fish species, such as this crocodile icefish,** live in waters that are very close to freezing and in some cases, below freezing point. To combat the risk of their body fluids freezing, they have blood proteins that act as antifreeze agents.

← **In areas where the winter ice** does not reach the seafloor, many species associated with hard substrates are able to flourish. The sponges and anemones here are only 20 feet (6 m) below the surface. They grow in cold, relatively tranquil waters, and are protected by their icy covering.

→ **In the Arctic, the close proximity** of continental landmasses has enabled the larger predators, such as the polar bear, and their attendant scavengers, such as the Arctic fox (*in the background*), to move out onto the sea ice. There, they go in search of the seals that are the polar bear's primary food. Polar bears can travel great distances, following seal migration routes, often drifting on large ice floes and swimming a long way to reach their prey. Male polar bears are larger than females. They can grow to a height of 5 feet 4 inches (1.6 m) at the shoulder, and weigh between 900 and 1600 pounds (410 to 720 kg). Polar bears are naturally wary of humans, but if they have been in contact with human settlements for any considerable length of time, they can become persistent and most unwelcome scavengers of trash.

The human impact

The oceans are highways, sources of food and mineral wealth—and, sadly, a site for dumping waste. Today, the importance of protecting the oceans for future generations is paramount. Potential benefits, such as clean energy and new drugs, remain, as yet, untapped.

Peoples of the sea: Inuit

The Inuit are one of the indigenous peoples of the Arctic and subarctic regions of Greenland, Alaska, Canada and far-eastern Russia (Siberia). They adapted over centuries to life in a polar climate by becoming almost totally reliant on the sea to supply their food and other needs. The lack of significant vegetation meant that caribou, seal, walrus, whale meat, whale blubber and fishes were their major food sources, as well as the source of raw materials for clothing, tents and boats. Seals were hunted from ice floes or from skin-covered kayaks. Harpooning whales required several people; it was done from larger craft known as umiaks. Families lived in hide tents in summer, and hunted caribou and other land animals with bows and arrows. They lived in snow-block igloos in winter, or in wood- or whalebone-frameworked houses dug into the ground and covered by stone or turf. During the twentieth century, contact with industrialized societies to the south brought major changes to Inuit life. The seminomadic life has now been largely abandoned, and many Inuit live in northern towns and cities. Out on the ice, snowmobiles have replaced traditional dogsleds, and hunting is done with a rifle.

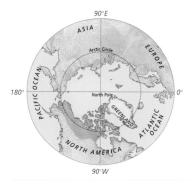

PEOPLES OF THE HIGH ARCTIC

Several closely related groups of people native to the high Arctic are found around the shores of the Arctic Ocean, and include the Inuit, Inupiat, Yupik and Alutiit. These are regional names that mean "the people" or "the real people." Today, the Arctic peoples of Canada and Greenland prefer to be known as Inuit (their distribution is shown, *left*), while the indigenous groups in Alaska refer to themselves as Eskimo, as do some of the indigenous peoples across the Bering Strait in Siberia. The origins of the peoples of the high Arctic are not clear, but archaeologists have found artefacts dating back around 3000 years on Umnak Island in the Aleutians, Alaska.

← **A hunter practices his harpoon** throw for spearing narwhals. Narwhals are small toothed whales with a characteristic long, spirally twisted tusk, that live in the coastal bays and inlets of Arctic Canada. Traditionally, they have been a highly prized source of food that could feed a family group for several weeks during the long winter months. Even a small narwhal can upset a canoe or drag it under the water, so a hunter's harpoon strike must be delivered with speed and deadly accuracy.

←← **An Inuit hunter may wait** many hours at seal breathing holes. Seals have a number of breathing holes and can remain submerged for 20 minutes between visits. The slightest disturbance near the hole will cause the seal to turn to an alternative one. As soon as a seal comes up for air, the hunter thrusts down a long pole with a hook, aiming to capture the seal before it can dive away.

← **This Inuit man wears** a traditional coat of caribou skin lined with fox or rabbit fur. Before the adoption of modern clothing, this man's trousers and boots would have been made from flexible and waterproof sealskins—perfect for life on the polar ice. The boots would have been further insulated by the addition of thick fur-felt soles.

→ **Although commercial whaling** is banned, the cultural importance of small-scale inshore whaling to indigenous coastal communities is recognized as vital. Here, in Alaska, a bowhead whale is flensed (stripped of skin and blubber).

Peoples of the sea: South Pacific Islanders

It is thought that the remote islands of the Pacific were first reached by humans in a succession of migrations from the mainland of southeast Asia and its long archipelago of islands. The first wave of exploration and settlement carried people to New Guinea, the New Hebrides and New Caledonia (Melanesia). Later generations moved northward to the Marshall and Gilbert Island groups (Micronesia), and then went eastward into the vastness of the Pacific and Polynesia. Beginning 4000 years ago, the Pacific migrations finished around 1000 years ago with the Maori settlement of New Zealand. The longest voyages took these early Pacific explorers as far east as the Marquesas Islands. From there, they ventured as far north as Hawaii, sometime between AD 450 and 600. They also sailed as far south as Easter Island. This remarkable dispersion over the Pacific was possible only because of the early development of large, seaworthy, sailing canoes, coupled with exceptional navigational skills.

→ **Three young Yakel islanders** in Vanuatu are being taught how to fish with a bow and arrow. Behind them is Yasur volcano.

↓ **These islanders** are using a beach net to catch small fishes. One end of the net is carried away from the beach by a small canoe which pulls the net out to its full extent. The ends of the net are hauled in and the catch brought to the beach.

KON-TIKI

The Kon-Tiki expedition was organized by the Norwegian scientist Thor Heyerdahl. He set out to demonstrate the feasibility of Tahiti and islands to the east being colonized by ancient explorers from the west coast of South America. Heyerdahl built a balsa wood raft from timber growing on the coast of Peru, and named it for an Inca deity. In 1947, he and five companions set off westward. After a voyage of 3½ months, covering approximately 5000 miles (8000 km) of ocean, they landed on Tahiti. The Kon-Tiki raft is now preserved in a museum in Oslo, Norway. However, subsequent anthropological research, supported by recent genetic studies of indigenous peoples of the Pacific, has negated Heyerdahl's hypothesis.

A group of children enjoy themselves on an outrigger canoe in the waters off Tsoi Islands, off Papua New Guinea. The projecting beam provides extra stability in the water during fishing, when the craft is heavily loaded, or in rough weather.

Local fishing

The seas have provided coastal communities with food and a livelihood for thousands of years. The earliest fishermen gathered buried bivalves from the beach, or picked out fishes and crustaceans from tide pools. Individual fishes could be caught from rocks and promontories by spearing, or with a baited hook and line. The next stage from gathering fishes from the shore was to build rock traps; these were artificial tide pools, trapping fishes as the tide receded. Rock traps continue to be used in many parts of the world. Other forms of inshore trapping that are still in use include increasingly elaborate net traps that funnel fishes into a series of conical nets, and the various forms of baited crab and lobster pots left on the seabed, marked by floats on the surface. Cast nets can be used from the shore, or from a small boat, to catch small shoaling fishes, and skimming nets are still used along the water's edge to catch shrimp. As nets and boats became larger over time, inshore fishing moved toward beach seine netting, a method still used in many European estuaries to catch migratory salmon. This has eventually led to powered inshore fishing boats, capable of seine netting or trawling as season and fish stocks dictate. Today, the shoreline remains the provider of food and a living for many people around the world.

↑ **The stilt fishermen of Sri Lanka** fish for sardines and other small fishes that move inshore at high tide. This style of fishing requires considerable endurance, since the poles can be reached only at low water. Once the tide is in, a fisherman has to wait until it has receded before he can return to shore. The pole sites are inherited and rarely pass out of a family group.

← **Vendors wash fishes** they have just bought at a fishing port in Cua Lo, Vietnam. With a coastline of over 1860 miles (3000 km), Vietnam has a great potential in sea products but needs huge investments to build ships and processing facilities in order to fully tap these resources. The poor capacity in the national fleet of offshore fishing vessels is said to be 30 to 40 years behind regional countries.

→ **These fishermen work** in the tidal waters of the Ganges delta, in the Sundarbans National Park in West Bengal, India. The park is one of the last refuges of the endangered Indian tiger so the fishermen use masks on the backs of their heads to protect themselves against tiger attack. The large, staring eyes are thought to mislead the tiger and prevent it attacking from the rear, since tigers attack by biting their prey across the back of the neck.

EARLY EXPLOITATION OF THE SEAS

Not all local fishing is connected with the harvesting of food items. Plato, nearly 2500 years ago, described the lucrative trade in sponges collected by divers from the Aegean. Until the advent of diving helmets, divers worked in the same way as those in antiquity. Small boats carried four or five divers offshore to water up to 100 feet (30 m) deep. Once an observer had spotted possible sponges, the divers were sent into the water. Divers usually worked naked and carried only a knife, and a large flat stone to take them to the seabed. Their time on the bottom depended on their stamina, but working for three to five minutes at 100 feet (30 m) was not uncommon.

→ **Cormorants are expert diving birds,** catching small fishes in coastal waters and estuaries. For many centuries, Chinese fishermen have used trained cormorants to catch fishes for them. The cormorants have a collar that prevents them from swallowing their catch, which they are trained to bring back to the boat. In order to catch a sufficient number of fishes, each fisherman has several cormorants.

Commercial fishing

Commercial fishing is a large-scale industrial process that employs increasingly efficient means of removing animals, mainly finfishes, from the sea. The basic commercial fishing gears are purse seining, trawling, gill netting and long-lining. Purse-seine netting uses a long net drawn around surface-living fishes such as tuna, sardines and herring. When the ends of the net are brought together, and the bottom hauled in, whole shoals of fishes are caught. Trawling uses a huge baglike net drawn over the bottom to catch bottom-living species such as flatfishes. To protect the mouth of the net, and to force fishes to swim off the bottom and into the oncoming net, trawls have heavy chains at the front that churn up the seabed. Gill nets are hung vertically, like an invisible mesh curtain in the water, to catch midwater fishes and are often 12 miles (20 km) long. Longlines are lines of baited hooks that are used in some areas to catch squid and tuna. They are seen as a less-damaging alternative to gill nets.

→ **Modern fishing fleets can stay at sea** for several months and traverse large distances. Factory ships (*pictured*) are now an integral part of the commercial fishing process, acting as a support ship for a group of smaller fishing boats.

THE COST OF THE CATCH

Despite many years of work on gear improvement, no fishing gear can be completely selective in the species that it catches. The accidental capture of non-target species, known as by-catch, is a serious problem in commercial fishing. For example, although haddock stocks in the North Sea are viable, fishing has been banned in areas where the effect of by-catches has threatened cod stocks. Gill nets are sometimes called the curtain of death, because these nets hang just below the surface and trap and drown thousands of dolphins, turtles and seabirds.

← **Pacific pollock are caught** in large numbers by trawlers in the rich waters of the Bering Sea.

→ **Women gut herring in England,** in 1921. North Sea herring stocks were once considered inexhaustible but overfishing caused a collapse.

TOP 25 FISHES CAUGHT (2001)

Name	Catch in tons (tonnes)
Anchoveta (Peruvian anchovy)	7,951,119 (7,213,077)
Alaskan (Walleye) pollock	3,457,388 (3,136,465)
Chilean jack mackerel	2,765,538 (2,508,834)
Atlantic herring	2,152,803 (1,952,975)
Japanese anchovy	2,024,413 (1,836,502)
Skipjack tuna	2,024,342 (1,836,438)
Blue whiting (Poutassou)	2,009,866 (1,823,305)
Chub mackerel	1,982,747 (1,798,704)
Capelin	1,841,873 (1,670,906)
Largehead hairtail	1,622,237 (1,471,657)
Yellowfin tuna	1,325,333 (1,202,312)
European pilchard (Sardine)	1,242,129 (1,126,832)
Atlantic cod	1,040,216 (943,661)
Atlantic mackeral	783,100 (710,411)
California pilchard	755,637 (685,497)
European anchovy	728,140 (660,552)
European sprat	713,661 (647,417)
Gulf menhaden	582,576 (528,500)
Japanese Spanish mackerel	576,244 (522,756)
Indian oil sardine	482,075 (437,328)
Kawakawa	466,821 (423,490)
Pacific herring	448,576 (406,938)
Pacific saury	414,663 (376,173)
Bigeye tuna	410,184 (372,110)
Pink (Humpback) salmon	397,908 (360,973)

Overfishing

At least 20 of the world's most important fisheries have disappeared since the mid-1970s, and many more have been so badly affected by overfishing that they are unlikely to recover. One of the best-known examples was the total collapse of the Grand Banks cod fishery in the northwest Atlantic in the early 1990s. A once-abundant cod stock became uneconomic to fish and stocks have not yet shown any signs of recovery. The problem of overfishing has been created by the doubling of the global demand for fishes and other marine animals over the last 30 years. This is the result of population growth in the developing countries and the need for cheap sources of protein. The only way wild stocks will survive as a viable and renewable resource will be to ensure that only a sustainable yield is taken. International controls need to regulate mesh sizes to ensure undersized fishes are not caught, and quotas should leave enough adults to breed, as well as leaving food for other marine animals.

→ **Over the last 40 years** there as been a dramatic decline in the numbers of sailfishes and other oceanic species. It is thought that this sharp decline began with the introduction of commercial long-lining.

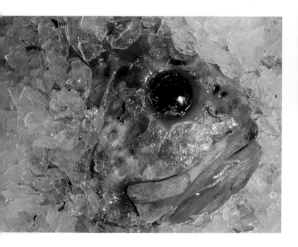

EXPLOITING THE DEEP

The best-known deep-water fishes that are caught and consumed by humans, are the orange roughy (*left*) and the black scabbard fish or espada. The orange roughy is found in the Atlantic, Pacific and Indian oceans but it is the stocks around New Zealand and, more recently, Australia, that have become a major deep sea fishery. Enormous numbers of these fishes have been landed annually since the fishery was first monitored in 1978. The sustainability of this fishery is debatable since the orange roughy is slow growing and takes years to become sexually mature. The Madeiran espada fishery is probably the oldest deep sea fishery, with records dating back to 1635. More recently, espada are now being fished off the northwest coast of Scotland.

↑ **Tuna are highly prized,** and sushi restaurants will sometimes pay thousands of dollars for a single fish. Tuna are mainly caught by drift nets that also kill marine mammals and seabirds. The high price paid for large tuna in good condition ensures the continuation of this form of fishing.

→ **Current estimates of fishing** activities, and the status of commercially fished stocks, show that year on year, fleets are catching fewer fishes for the same, or greater, fishing effort. It is also apparent that many commercial stocks have been so overexploited that it will take many years for them to recover, but only if some form of regulation, such as the limitation to maximum sustainable yields, can be enforced.

FISHERIES IN DECLINE

Species	Peak year	Peak catch*	1992 catch*	Decline*	Percent change
Pacific herring	1964	0.7 (0.64)	0.2 (0.18)	0.5 (0.45)	-71%
Atlantic herring	1966	4.1 (3.7)	1.5 (1.4)	2.6 (2.36)	-63%
Atlantic cod	1968	3.9 (3.5)	1.2 (1.1)	2.7 (2.45)	-69%
South African Pilchard	1968	1.7 (1.5)	0.1 (0.09)	1.6 (1.45)	-94%
Haddock	1969	1 (0.91)	0.2 (0.18)	0.8 (0.73)	-80%
Peruvian anchovy	1970	13.1 (11.9)	5.5 (5.0)	7.6 (6.89)	-58%
Polar cod	1972	0.35 (0.32)	0.02 (0.02)	0.33 (0.29)	-94%
Cape hake	1972	1.1 (1.0)	0.2 (0.18)	0.9 (0.81)	-82%
Silver hake	1973	0.43 (0.39)	0.05 (0.05)	0.38 (0.34)	-88%
Greater yellow croaker	1974	0.2 (0.18)	0.04 (0.04)	0.16 (0.15)	-80%
Atlantic redfish	1976	0.7 (0.6)	0.3 (0.27)	0.4 (0.36)	-57%
Cape horse mackerel	1977	0.7 (0.6)	0.4 (0.36)	0.3 (0.27)	-46%
Chub mackerel	1978	3.4 (3.1)	0.9 (0.81)	2.5 (2.27)	-74%
Blue whiting	1980	1.1 (1.0)	0.5 (0.45)	0.6 (0.54)	-55%
South American Pilchard	1985	6.5 (5.9)	3.1 (2.81)	3.4 (3.08)	-52%
Alaska pollock	1986	6.8 (6.2)	0.5 (.45)	6.3 (5.71)	-93%
North Pacific hake	1987	0.3 (0.27)	0.06 (0.05)	0.24 (0.21)	-80%
Japanese pilchard	1988	5.4 (4.9)	2.5 (2.27)	2.9 (2.63)	-54%
Totals		**51.48 (46.61)**	**17.27 (15.71)**	**34.21 (30.99)**	**-58%**

* millions of tons (millions of tonnes)

Fish farming

Fish farming, or aquaculture, has a long history that can be traced back to 2500 BC; Egyptian temple reliefs from that time depict the harvesting of fishes from ponds. Today, marine fish farming plays an increasingly important role in meeting the growing demand for food as the world's population rises. It is also helping overcome the reliance on wild stocks, many of which are now threatened, or have already collapsed. There are commercial advantages in farming shellfishes and fishes compared to catching them in the wild. Species of a reasonably uniform size, weight and color can be produced to suit the retailer or consumer, and can be supplied at regular times, in predictable quantities. Some species are now produced all year round.

Although most marine fishes are farmed in cages, migratory species are "ranched." These species are raised in hatcheries and released into the wild with the expectation they will return to home waters after feeding at sea. Currently, the main marine fishes raised on fish farms are tuna, sea bass, sea bream, mullet, flatfishes, cod and species of salmon.

STAGES IN AQUACULTURE PRODUCTION

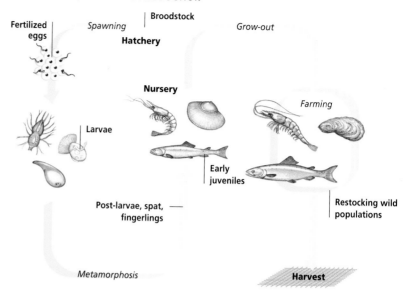

AQUACULTURE
Adult broodstock provide eggs and sperm that are artificially fertilized to ensure the maximum production of larvae. At the larval stage, the fishes are grown under carefully controlled conditions in a hatchery. When strong enough, they are moved to large growing ponds or underwater cages. At this stage, some will be selected to maintain the broodstock. The rest are grown for harvesting or restocking programs.

← **Fishes are harvested** from onshore ponds either by draining the ponds or by netting. Although netting is the more labor-intensive of the two, it allows for size selection and causes less damage to fishes.

↑ **Carp are the most widely farmed** freshwater fishes. The majority of fishes farmed are freshwater species.

→ **Atlantic salmon are reared** intensively in cages in sheltered inlets and bays. There are concerns about the environmental impact of these farms relating to escaping fishes, seabed fouling and the increased threat of disease among wild stocks of salmon.

↓ **Rainbow trout are usually grown** in freshwater but can be acclimated to life in saltwater cages. Rearing in seawater improves their flavor and increases their value when they are marketed as salmon trout.

Shellfish farming

The term shellfish covers three large groups of marine invertebrates—mollusks, crustaceans and echinoderms. Their cultivation has a long history, with records of intertidal oyster farming in Japan as far back as 2000 BC. Later, Aristotle and Pliny described Greek and Roman methods of bivalve farming around the shores of the Mediterranean. The scale of modern commercial shellfish farming is shown by statistics published by the Food and Agriculture Organization of the United Nations. It estimated that in 2001, commercial farming of crustaceans, mainly shrimp, produced 1.65 million tons (1.5 million t) of shellfishes worth 8 billion US dollars. The production of mollusks was even greater, yielding 7.5 million tons (6.8 million t), worth 9.28 billion US dollars. The most important species are the bivalve mollusks, principally oysters, mussels, scallops and clams, as well as a small number of gastropods such as abalone, whelk and conch. The main crustacean species farmed on a commercial scale is shrimp, and there is some interest in developing lobster farming on a large scale. The advent of rapid air transport has opened up a global market for commercially farmed shellfishes that were once limited in their distribution range by their perishability.

↑ **Mussels are often farmed** using rope culture. Specially frayed ropes are suspended over spawning mussel beds so they are "seeded" with young mussels. The ropes are often moved to growing areas where water quality is best.

→ **The intensive cultivation** of introduced Pacific and native flat oysters is a major part of the economy in several areas on the coast of France. Here, Pacific oysters are being prepared for transfer to bags set out on trestles on the shore.

TOTAL SHELLFISH PRODUCTION (2001)

Species group	Aquaculture tons (tonnes)	Wild caught tons (tonnes)	Total tons (tonnes)
Oysters	4,638,349 (4,207,818)	219,377 (199,015)	4,857,727 (4,406,833)
Shrimp and prawns	1,400,907 (1,270,875)	3,252,754 (2,950,834)	4,653,661 (4,221,709)
Squid and cuttlefish	17 (16)	3,689,265 (3,346,828)	3,689,283 (3,346,844)
Clams and cockles	3,427,130 (3,109,024)	89,227 (80,945)	3,516,357 (3,189,969)
Other mollusks	1,482,355 (1,344,763)	1,662,138 (1,507,859)	3,144,493 (2,852,622)
Mussels	1,510,869 (1,370,631)	283,642 (257,315)	1,794,512 (1,627,946)
Other crustaceans	39,989 (36,278)	1,548,274 (1,404,564)	1,588,264 (1,440,842)
Crabs and sea spiders	181,035 (164,232)	1,208,548 (1,096,371)	1,389,584 (1,260,603)
Scallops	142,438 (129,217)	774,405 (702,525)	916,843 (831,742)
Lobsters	38 (35)	248,171 (225,136)	248,209 (225,171)
Abalones and conches	5980 (5425)	133,334 (120,958)	139,314 (126,383)
Krill	0 (0)	108,297 (98,245)	108,297 (98,245)
King crabs and squat lobsters	0 (0)	49,570 (44,969)	49,570 (44,969)
Total annual production 2001	**12,874,107 (11,638,314)**	**13,222,012 (12,035,564)**	**26,096,119 (23,673,878)**

SHELLFISH BY-PRODUCTS

Processing shellfishes on an industrial scale produces major waste disposal problems. However, a number of uses for waste shells have been devised. Cleaned scallop shells are used as seafood dishes in catering, while cockle and mussel shells have been rediscovered as thermal and sound-insulating materials in buildings. Crustacean shells are now the source of chitosan—an alternative to wax, to coat fruit and other food.

↑ **Japanese oyster growers** mainly cultivate the Pacific oyster. This species grows rapidly in the mesh boxes suspended beneath each float, but regular checking is required to remove seaweed and organisms that restrict water flow over the oysters.

← **This chart shows** the total worldwide shellfish production for 2001. Many species of shellfish are cultivated for ever-increasing human consumption.

Pearling

Pearls are produced naturally by bivalves that cover irritants inside the shell with shiny layers of nacre (mother-of-pearl) to isolate the problem. Pearls are found in both freshwater and saltwater species, and come in all colors, including black. Saltwater pearls are considered to be superior in color and form, with the Indian rose-pink variety the most prized. The largest pearls are usually irregular in shape and known as baroque pearls. The largest baroque pearl recorded weighed some 3⅓ ounces (93 g). Today, the main sources of high-quality natural pearls are the beds of pearl oysters in the Persian Gulf off Qatar, and the Gulf of Mannar in the Indian Ocean. There are Chinese records of culturing pearls in freshwater mussels dating back 3000 years, but modern saltwater-pearl culture was developed in Japan in the 1890s. Japan and Australia are now the world's major producers of cultured pearls.

WORLD PEARL TRADING 2000–2003			
Exporters	**Trade value (US$)**	**Importers**	**Trade value (US$)**
Japan	$1,006,468,432	USA	$1,017,816,704
China, Hong Kong SAR	$848,891,104	Japan	$813,853,920
Australia	$642,072,992	China, Hong Kong SAR	$700,972,320
French Polynesia	$410,372,952	Switzerland	$219,762,424
Switzerland	$220,183,720	Germany	$178,547,520
Others	$746,853,811	Others	$743,360,108
Total export:	**$3,874,843,008**	**Total import:**	**$3,674,313,024**

⤒ **Female pearl divers sort** through their catch of pearl oysters collected from the offshore beds in Ago Bay, Japan, in the 1950s.

↑ **A diver inspects cages** containing pearl oysters in the warm waters of Thailand. Keeping the oysters in hanging cages improves their growth rate.

→ **A worker sorts through cleaned shells.** It is possible to produce three or four large, high quality pearls from each animal.

← **Japanese pearl-production beds** grow the "seeded" pearl oysters in wire nets, hung from floating rafts in shallow water.

Nacre layer
Sand grain

Natural pearl

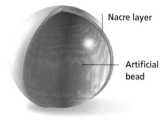

Nacre layer
Artificial bead

Cultured pearl

Whaling

European whaling can be traced backed to the thirteenth century, but intensive whaling began in the eighteenth century, stimulated by the demand for lamp oil and baleen "whalebone" for corsets. The replacement of these items by mineral oil and steel in the 1850s, however, coincided with the invention of margarine and the need for cheap food to feed industrial workers. Hence, whaling continued well into the twentieth century. By the 1960s, the decline in the great whales forced the introduction of quotas, beginning in 1965. A complete ban was placed on commercial whaling in 1985. Japan, Iceland and Norway have continued to hunt smaller whales under the guise of "scientific whaling."

← **The Faroe Islands are** a semi-autonomous region of Denmark, a signatory to the 1985 moratorium. However, small toothed pilot whales are still driven ashore and killed.

→ **With the invention of high-powered** harpoon guns, explosive charges and electrical stunning, a whale will die in a much quicker time than in the early days of whaling.

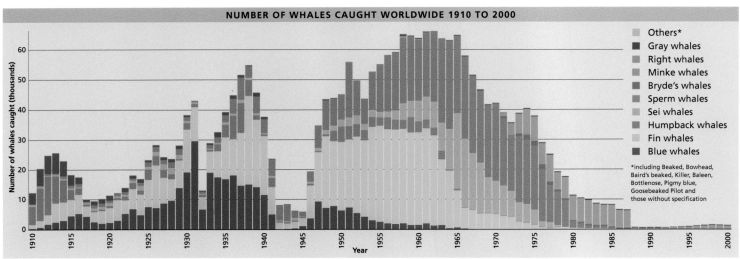

NUMBER OF WHALES CAUGHT WORLDWIDE 1910 TO 2000

Legend:
- Others*
- Gray whales
- Right whales
- Minke whales
- Bryde's whales
- Sperm whales
- Sei whales
- Humpback whales
- Fin whales
- Blue whales

*including Beaked, Bowhead, Baird's beaked, Killer, Baleen, Bottlenose, Pigmy blue, Goosebeaked Pilot and those without specification

y-axis: Number of whales caught (thousands)
x-axis: Year (1910 to 2000)

WHALE NUMBERS	
Whale	Estimated no.
Minke whales	935,000
Blue whales	400–1400
Fin whales	47,300
Gray whales	26,400
Bowhead whales	8000
Humpback whales	21,500
Pilot whales	780,000

↓ **Icelandic whalers flense a sperm whale.** Once killed, the whale is inflated with compressed air to keep it afloat while it is hauled onto the factory ship. Inflation also helps to separate the skin and blubber from the muscles. Iceland continues to hunt and process whales for food, claiming that its whalers hunt only from sustainable whale stocks. Iceland resigned from the International Whaling Commission (IWC) in 1992, after being subjected to sustained pressure from other commission members.

WHALING IN THE TWENTIETH CENTURY

Whale catches rose steadily during the twentieth century (see chart, *left*) as whaling fleets became larger and could stay at sea for extended periods, processing catches on factory ships. The development of radio communications meant that spotter planes used to locate animals became commonplace. The increasing catches reflected the greater efficiencies in hunting, and disguised the catastrophic decline in numbers of many of the great whales, whose low reproductive output made them especially vulnerable to over-exploitation. The International Whaling Commission introduced protection for the blue, gray and humpback whales from 1965 onward, but stocks were already too small to be worth hunting. Sperm, minke, fin and sei whales continued to be caught, but in declining numbers up to the 1985 international moratorium on whaling. Today, even the great whales are starting to show some recovery in numbers.

Oil and gas

Oil and gas were formed through the geological transformation of countless plants and animals that lived hundreds of millions of years ago. The remains of these lifeforms sank to the seabed where, covered with mud and sand, they were compressed by overlying rocks and gradually transformed over millions of years into crude oil and natural gases. Oil and gases tend to move upward through porous rock until they reach an impermeable layer known as caprock. They remain trapped until the caprock is penetrated by a drill.

Offshore oil and gas production centers mainly around the Middle East, Central and South America, and the North Sea. Substantial reserves have been found off the coasts of Nigeria, Egypt and Indonesia. In 2002, 76.33 billion barrels of oil were produced, with the figure set to rise as human demand becomes stronger.

OIL AND GAS LOCATIONS

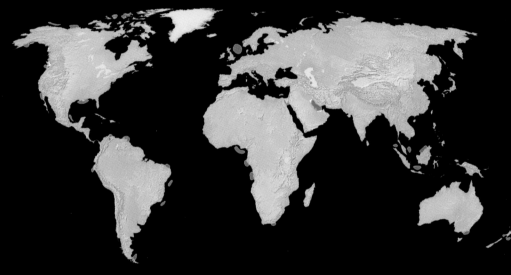

↑ **The main areas of offshore** oil and gas production are shown. This multibillion-dollar industry globally employs a vast workforce.

← **The flare stack on this North Sea** oil rig burns off excess gases. Not all gases released from the drilled rocks can be used for natural gas supplies, or liquefied for the petrochemical industry.

ENVIRONMENTAL IMPACT

To locate oil and gas deposits, sound waves are fired at the seabed and the resultant signals are processed to assess the geology of the area. These seismic surveys pose a danger to marine life. Mammals may suffer hearing damage, and fishes with gas bladders can be killed in large numbers by the high energy emitted. Survey impacts are usually short-lived, but production can last many years. The major impact of drilling is caused by the dumping of a toxic mixture of detergents and heavy metals called drilling muds.

↑ **A typical offshore oil platform.** Several hundred workers live on these platforms for weeks at a time, working to pump out millions of barrels of oil per day. Most platforms stay operational for 25 years.

← **These oil rigs in Texas, USA,** were first erected in 1919 to extract oil from nearshore deposits. The rigs tap into a small part of the oil fields that extend into the deep waters off the Gulf of Mexico.

Salt and other minerals

Between 3.3 and 3.7 percent by weight of seawater is made up of dissolved salts. These salts have been extracted from the sea for thousands of years; this is done by trapping seawater in shallow pools and waiting for the water to evaporate, leaving behind the crystallized sea salt. Although other sources of salt, such as mines and brine wells, now contribute to salt production, especially in inland areas, salt pans are still one of the main sources of sea salts. However, the process today is carried out on an industrial scale.

Seawater is also an important source of magnesium, since it is the third most abundant element in seawater (after sodium and chloride). Magnesium is used to produce strong, light components for aircraft and precision engineering. It is extracted from concentrated brine by electrochemical means; about half the world's production of magnesium metal is from seawater. A variety of other minerals are collected from the seafloor. Apart from sand and gravel for construction, phosphates in the form of deposits of phosphatite are abundant in shallow waters off Florida, the Pacific coast of South America and the Atlantic coast of west Africa. In the deep sea, manganese nodules and vent deposits are sources of mineral wealth.

↓ **These ancient salt pans** on the Mediterranean island of Gozo were made by building small walls on the flat rock. Seawater is pumped into the hollows and left to evaporate in the heat of the Sun. The pans have been in use for at least several thousand years, producing one of the more valuable commodities of the ancient world.

A third of the world's table salt still comes from the sea. It is often held in giant stacks before processing. Seawater also provides a number of useful bulk chemicals with a variety of uses: Concentrated brine can be treated to yield magnesium and other metals; potassium salts are used as fertilizers and chemical reagents; bromine is used in medicines, plastics and the petrochemical industry; gypsum from the sea is used to make building materials.

↓ **Modern sea salt production** is carried out on an industrial scale. By regulating the amount of water in the ponds and transferring the brine to another pond at just the right point, it is possible to separate the major salts. In modern salt works, the composition of the brine is monitored by sensors, and powerful pumps move brine between ponds automatically.

MANGANESE NODULES

Manganese nodules (*left*) were first discovered on the Pacific seafloor by the British vessel HMS *Challenger* in 1874. These distinct, rounded lumps have now been found scattered around the abyssal plains across all the world's oceans. They are rich in iron, manganese, copper, nickel and cobalt. It is estimated that the deposits found in the Pacific alone exceed 17.6 billion tons (16 billion tonnes) and are the equivalent of 2000 years of production at present usage. The extreme depth of the nodules, lying 12,000 feet (4000 m) beneath the surface, makes their recovery uneconomic at present, but this will change when land sources are exhausted.

Desalination

Desalination is the process of converting saltwater into freshwater. It is especially useful in arid areas, or where human demand outstrips local freshwater resources. Today, desalination provides water for many parts of the world, including California and Florida in the USA, the Arabian peninsula and Hong Kong. There are several methods of desalination, the oldest of which is distillation. Saltwater is boiled and the resulting vapor is cooled to condense into freshwater. This process is expensive, however, because it consumes large amounts of energy. It also produces a highly concentrated salt solution that can cause severe environmental damage when discharged into coastal waters. Reverse osmosis is a much more energy-efficient method of desalination and is replacing distillation in many areas. Apart from a reduction in environmental impact compared to distillation, an advantage of this method is that it can be scaled down to suit the needs of individual users and small communities.

↓ **Reverse osmosis desalination** plants consist of thousands of individual units. They filter out salt and impurities by passing seawater through a membrane. This plant in Tampa Bay is the largest reverse osmosis plant in the USA. It provides the region with 10 percent of its drinking water.

↗ **Desalination plants are constructed** in coastal areas where there is little or no freshwater supply from rivers, such as coastal towns in desert areas. Biological filters remove any organisms from the intake water. The desalination plant then removes the salt and other minerals from the seawater.

↓ **If freshwater and saltwater** are separated by a semi-permeable membrane, freshwater is drawn into the saltwater (osmosis). Applying pressure to the saltwater will stop the flow (osmotic equilibrium). Greater pressure will force pure water back out of the saltwater (reverse osmosis).

OSMOSIS

Freshwater moves across | Membrane | Salt molecules are too large to cross | Saltwater

OSMOTIC EQUILIBRIUM

Freshwater | Head of saltwater stops osmosis

REVERSE OSMOSIS

Freshwater | Pressure forces water through the membrane leaving salt behind

WATER USAGE

Domestic use of water has increased many times over since the beginning of the twentieth century, as the world's population increases and economic changes have brought about increasing affluence. However, direct human consumption is only a small part of global water consumption; 70 percent of water usage is for irrigation to feed the rising population.

Modern desalination plants have to take account of increasingly stringent environmental controls on their saline discharges. Many distillation plants, such as this one at Lake Mead, Nevada, USA, now have

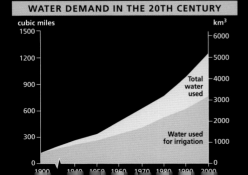

WATER DEMAND IN THE 20TH CENTURY

cubic miles			km³
1500			6000
1200			5000
		Total water used	4000
900			3000
600			2000
		Water used for irrigation	
300			1000
0			0

1900 1940 1950 1960 1970 1980 1990 2000

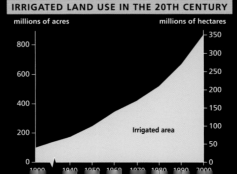

IRRIGATED LAND USE IN THE 20TH CENTURY

millions of acres		millions of hectares
		350
800		300
		250
600		200
400		150
	Irrigated area	100
200		50
0		0

1900 1940 1950 1960 1970 1980 1990 2000

Tidal power

Tidal power was first harnessed hundreds of years ago for tasks such as milling flour, sawing timber and grinding gunpowder. In the early twentieth century, it was harnessed to generate electricity. The first tidal power station was built in 1913 on the Dee estuary in England, and the first successful large-scale plant was constructed by French engineers between 1961 and 1967. It spans the estuary of the Rance River that flows into the Gulf of Saint-Malo, Brittany. There, a barrage equipped with reversible turbines allows the station to operate during every stage of the tidal cycle. The tidal flow's power is therefore harnessed in both directions—from the sea to the tidal basin on the flood, and from the basin to the sea on the ebb. Rance has 24 power units of 10,000 kilowatts each; about seven-eighths of the power is produced on the more controllable ebb flow. In Canada, tidal power stations have been built to exploit the tides in the Bay of Fundy, which experiences some of the biggest tidal ranges in the world. The tidal range reaches more than 49 feet (15 m). Large amounts of power can be generated from tides in favorable locations and conditions, but power is intermittent and varies with the seasons.

↓ **The restored tide mill** at Birlot, France, was built between 1634 and 1637 to grind wheat. The large tides on the Brittany coast are used to fill a 5 acre (2 ha) pond behind a 460 foot (140 m) dyke.

→ **The La Rance tidal barrage,** France, has a maximum tidal range of 44 feet (13.5 m). It is 2461 feet (750 m) long and creates a lake of 8½ square miles (22 km²). The generators can produce 240 million watts.

ENVIRONMENTAL COSTS

There are major environmental and ecological consequences associated with the building of a tidal generating barrage. While every site has unique characteristics and raises specific considerations, some consequences are predictable and common to all sites. A change in water level, and possible flooding, affect vegetation around the coast, impacting on aquatic and shoreline ecosystems. The quality of water in the basin, or estuary, is also affected. Sediment levels change, altering the turbidity of the water and, in turn, affecting animals such as fishes and birds that live in it or depend upon it. Fishes, in particular the important migratory species such as salmons, sea trout and eels, will be harmed or killed by turbines unless provision is made for them to pass through the barrage. Each environmental and ecological consequence affects the types of wading birds and other birds in an area. When food sources are under threat, they must migrate to an area more favorable to their survival.

At Eling mill, England, there is a double high water in the tidal cycle. **1.** During the first high tide, the pond and the estuary are at the same level so there is no flow. Between the first and second high waters, the sea gate remains shut, holding the water in the pond. **2.** Once the tide has started to drop, the gate is opened but the wheel does not turn because of the drag of the water on the estuary side. **3.** As the tide drops further, the wheel can turn and there is power to turn the millstones. **4.** The mill continues operating during the low water until either the pond is exhausted or the tide turns.

A tide mill has been standing at Eling, near Southampton, England, for more than 900 years. It is believed that one may have been on the site since Roman times. For many years, Eling mill was operated as a commercial enterprise, with grain being carried by sea. In the 1930s, the mill was abandoned. In 1975, it was bought for restoration and produces flour once again. When working at full capacity, up to 4½ tons (4.1 t) of flour a day can be milled.

1. HIGH TIDE

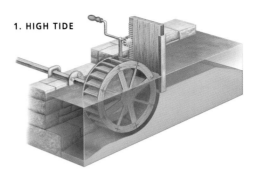

2. 1½ HOURS AFTER SECOND HIGH WATER

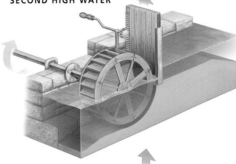

3. EBB TIDE

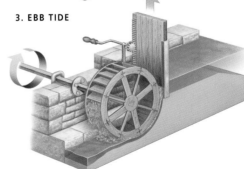

4. LOW WATER

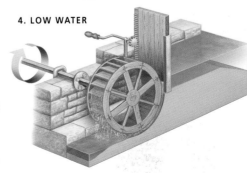

Wind and wave energy

Interest in the generation of energy, mainly electricity, from renewable resources began in the 1970s when the oil crises of 1973 and 1979 focused attention on supplies of fossil fuels. Development of new energy-generation technologies was further stimulated by the recognition that the emission of large amounts of carbon dioxide was affecting Earth's climate. Renewable energy sources are viewed as a more acceptable long-term energy source than the only other major alternative to fossil fuels—nuclear power. A number of technologies have been developed to exploit the energy that can be harnessed from the sea. These are wave-powered generators, tidal barrages, underwater turbines, ocean thermal gradients and coastal or offshore wind farms. Current estimates of the generating capacity of the oceans indicate that there is a capacity in excess of 450 gigawatts, and a potential maximum of more than 10 times this amount. At full capacity, the power generated from the oceans would more than cover the world's power needs, which are currently thought to be about 3,500 gigawatts.

↓ **The world's first wave** turbine power station is on the Scottish island of Islay. It generates 500 kilowatts of electricity by harnessing the power of water flow.

↗ **The *Osprey*, launched in 1995,** is the world's first commercial offshore wave-energy generator. The structure produces 2 megawatts of electricity.

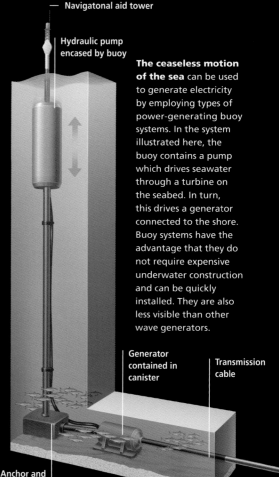

— Navigatonal aid tower

Hydraulic pump encased by buoy

The ceaseless motion of the sea can be used to generate electricity by employing types of power-generating buoy systems. In the system illustrated here, the buoy contains a pump which drives seawater through a turbine on the seabed. In turn, this drives a generator connected to the shore. Buoy systems have the advantage that they do not require expensive underwater construction and can be quickly installed. They are also less visible than other wave generators.

Generator contained in canister

Transmission cable

Anchor and turbine

Coastal and offshore sites are favored locations for wind farms. These places provide unimpeded exposure to winds from all directions and access to water to cool the larger generators. When compared to burning coal or oil, utilizing a single 660 kilowatt wind turbine will, over the course of a year, avoid emissions of 1100 tons (998 t) of carbon dioxide (the main greenhouse gas), 6 tons (5.5 t) of sulfur dioxide (the cause of acid rain), and 4 tons (3.6 t) of nitrogen oxides, one of the main causes of ground-level atmospheric pollution.

The top of a wind turbine rotates so that the blades face into the wind, causing them to turn at a relatively slow speed. The hub of a turbine is connected to a gearbox that increases the speed of rotation of the shaft connected to the generator. Large wind turbines are used to generate electricity for regional or national utilities and can produce between 50 and 750 kilowatts per turbine.

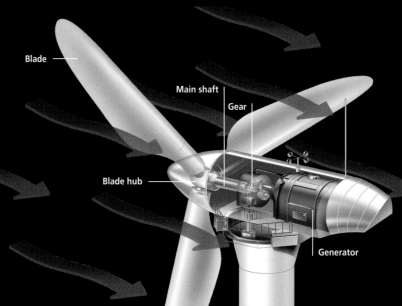

Blade

Main shaft

Gear

Blade hub

Generator

Habitat destruction

The loss of marine habitats is an ever-present and growing problem that largely goes unnoticed; it is out of sight and out of mind. The collapse of many fisheries has forced fishing fleets to sail into deeper water and to expend much greater effort in catching the remaining stocks. This has not only damaged fish stocks, but trawling itself causes major damage to the seabed. The heavy gear that is dragged across the seafloor kills animals unable to move and catches others (by-catch) that are later discarded dead. Fisheries, such as the North American gulf shrimp fishery, discard up to 90 percent of their catches. In some areas, the seabed is constantly turned over so that there is no chance of recovery. At the margins of the sea, land reclamation of estuarine mudflats, saltmarshes and mangroves for industries and, ironically, aquaculture, continues to destroy unique marine communities and damage coastal fish populations by removing important nursery grounds. There are efforts to repair some damage by replanting mangroves on abandoned shrimp farms and allowing the sea to breach old sea defenses to restore temperate saltmarshes.

→ **The impact of rig accidents** is not confined to the area of the seabed where they come to rest. Oil and other chemicals will leach out over decades, contaminating a large area around the wreck.

← **Trawl nets have "tickler"** chains at the mouth to stir up the seabed ahead of the net so fishes swim off the bottom and into the net.

LAND RECLAMATION

There have been many large-scale land-reclamation schemes. The reclamation of much of the Netherlands since the seventeeth century, culminating in the creation of IJsselmeer (formerly Zuiderzee), is the best-known and biggest of these schemes. However, in the Netherlands and other parts of the world that have extensive coastal defenses, the need to continually raise the height of these barriers in response to the rise in sea levels has prompted a reconsideration of coastal management.

← **West Kowloon reclamation** scheme in Hong Kong, is one of a number of schemes in an economically vibrant area where the room for expansion of population centers is restricted. This scheme has reclaimed 100 acres (40 h) from the harbor.

THREATENED HABITATS		
Habitat type	**Threat**	**Quantifiable loss**
Shelf-seas seabed	Commercial trawling	Estimated powered trawler coverage: 1.4 billion square miles (3.6 billion km^2) per annum.
Seagrass beds	Land reclamation	15% world's seagrass beds lost 1993–2003
Cold-water corals	Commercial trawling	4.4 tons (4 t) per day removed in NE Atlantic
Warm-water corals	Bleaching, dynamiting and trawling	95% of all reefs show some form of damage 30% of known reefs seriously damaged in 2003, estimated to be 60% by 2030.
Mangroves	Aquaculture, coastal development	Only 50% of world's mangroves still intact in 2000. Estimated loss rate 2–8% of total area and 95% loss predicted by 2040.
Temperate wetlands	Land reclamation	50% of world's total area lost since 1900. Estimates suggest 31% of salt marshes and 37% of coastal lagoons will be lost by 2050.

Human impact: estuaries

Estuaries have been a natural focus for human settlement because of a unique combination of geographical factors. Many are found on coastal plains where there is flat land to allow settlements to expand, rivers to provide transport routes inland, and proximity to the sea that allows access to long-distance trade routes. During the late nineteenth and early twentieth century, estuaries became important industrial sites because of their transport links via the sea. Coal, metal ores and later, oil, have been the main materials that supported heavy industries associated with estuaries. Until the advent of pollution control measures, estuaries provided an easy route for the disposal of waste products from industry. In the twentieth century and today, the industrialization of estuaries has continued, most prominently with the siting of electricity plants.

CLEANING CAPACITY

The human impact on estuaries through industry and land reclamation has resulted in them becoming the most polluted of all marine habitats. However, estuaries have some capacity to clean themselves. Fine particles brought down in suspension by rivers, tend to absorb a number of pollutants from the water, especially metals. When the particles are deposited as fine muds, many of the pollutants are removed from the water column. Bacteria in muds also bind to metals and lock them into the sediments. Bacteria can also break down organic chemicals, including some petrochemical wastes. Plant life in estuaries removes nitrogen compounds from sewage and other sources, as well as scavenging out any number of metals from the water.

↑ **Land use around Medway estuary,** in the United Kingdom: river (dark blue); salt marsh and intertidal land (mid-blue); pasture (green); arable land (brown); beaches or mudflats (pale yellow); low (gray) and high density (black) built-up areas.

↖ **Hot water is as destructive** as the many toxic chemicals discharged into the sea. Heating disrupts the lifecycles of many marine organisms.

↖ **Nuclear power-station reactors** are relatively inefficient—some 60 percent of the heat energy generated is discharged into an estuary or the sea.

← **Heavy industry on estuary shores** reduces both local water and air quality.

Coastal pollution

Pollution can be defined as the introduction of substances, material or unwanted heat energy from human activities, that adversely affect an ecosystem. The vastness of the oceans once disguised the growing problems caused by the use of the seas as a dumping ground. However, marine pollution has not only become more visible as plastic wastes ensnare animals and litter the beaches, but more insidious problems, such as the chemical contamination of marine food chains, have been uncovered by improving analytical methods. In coastal waters, there are two main sources of pollution. The majority of inshore pollution is from land run-off, while the remainder is dumped or pumped into the sea. Various initiatives around the world have started to tackle the problem and there have been some successes. The Regional Seas Program to clean up the Mediterranean commenced in 1975; it is no longer one of the most polluted seas in the world.

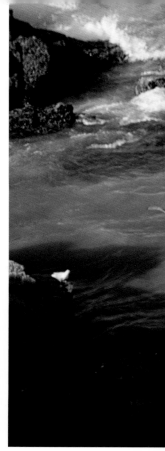

↓ **Oiled seabirds are one of the most** emotive images associated with oil spills. Diving birds are especially vulnerable to oil contamination since the oil clogs their feathers, which provide insulation and buoyancy in the water. For every oiled seabird that survives long enough to be washed ashore, many more have died at sea.

↑ **The development of plastics** has been a key component in shaping the modern world, but their benefits have come at environmental cost. Discarded plastic packaging, fishing nets and ropes take many years to break up. In the meantime, marine animals can become entangled in them.

→ **Each year, enormous** amounts of plastics, ropes, fishing nets, glass and metal are dumped in the sea. Some trash will be broken down by wave action, but most is non-biodegradable and will persist in the marine environment for tens of years.

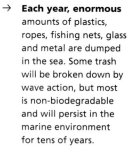

OIL SPILLS

Crude oil is one of the most widespread pollutants in the marine environment. Release occurs by spillages at refineries, tanker accidents and blowouts of pipelines and drilling rigs. There are also natural oil discharges into the sea from deep sea seeps. The effects of an oil spillage can be devastating. Aquatic birds and mammals will die slowly from exposure, starvation and the absorption of toxic chemicals when covered in crude oil. Fishes and other marine life will be affected by the toxic components that are soluble in water. Shore life will be smothered by a tarry coating. In the past, detergents and steam cleaning have been used to clean up oiled shores, but these have been shown to be worse than leaving the shore alone. Crude oil is biodegradable and in most cases, the containment and removal of the bulk of the oil where it comes ashore is employed. Experiments have been conducted with oil-eating bacteria to speed up the natural breakdown of oil.

↑ **Coastal outfalls** not only litter beaches with unpleasant material, but also make the water unsafe for swimming and other water sports. The large amounts of nutrients in the sewage can also produce excessive, sometimes harmful, algal blooms.

→ **Workers cleaning a shore** after a major oil spill use a suction pipe to scoop up a sticky mess of oil, sand and seawater. Oil slicks not only smell bad, but often contain large amounts of volatile chemicals that can cause headaches and nausea after a few minutes of exposure.

Ocean dumping

The disposal of wastes at sea has been based on the belief that the immense size of the oceans will dilute any contaminant to vanishingly small levels. However, while present knowledge has shown that materials that have been put into coastal waters adversely affect marine communities, we still have little knowledge of how the deep sea might react to the addition of large amounts of wastes. To date, only the "106-mile experiment," off New York, has shown that the deep sea ecosystem is altered by sewage sludge dumping, and that the effects spread beyond the boundaries of the dump site. In the past, the deep sea has been used for the disposal of large amounts of conventional munitions, smaller quantities of chemical agents and radioactive wastes from civil and military sources, including obsolete nuclear submarines.

→ **The steel in this motorbike** will take at least 50 years to rust away and the aluminum engine at least 200 years to dissolve. Oil and petrol residues will harm animals in the vicinity and some chemicals will pass up the food chain.

→→ **Discarded plastic items lie** at the bottom of the Mediterranean Sea. Apart from the damage it causes marine environments, this garbage will take centuries to break down.

↓ **Here, a glass bottle is home** to a tropical mantis shrimp. Glass is non-biodegradable and chemically stable in seawater. Theoretically, it can remain unaltered for an unlimited period. It merges into the environment only by being broken down.

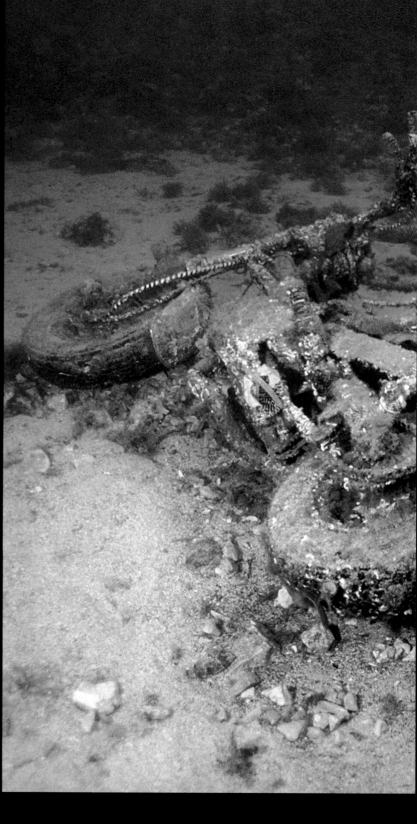

→ **The International Convention** for the Prevention of Pollution from Ships (MARPOL) is a combination of two treaties adopted in 1973 and 1978, and regularly amended since its inception. The initial regulations were drafted to deal with the problems associated with oil dumping and other types of pollution.

MARPOL LIMITS OF DISCHARGE	
Dumped material	Distance from coast nautical miles
Treated garbage	3
Treated sewage	4
Some noxious liquids*	12
Untreated garbage	12
Untreated sewage	12
Oil discharge**	50

* No discharge in Black Sea or Baltic Sea

**No discharge in Mediterranean Sea, Baltic Sea, Red Sea, Black Sea or Persian Gulf

Global warming

Earth has a natural temperature-control system sometimes referred to as the Gaia phenomenon. The functioning of this system depends on a critical balance between the main atmospheric gases. The most important of these are the greenhouse gases, so called because they act like glass in a greenhouse, preventing radiation from being reflected back into space. The trapped infrared radiation warms the atmosphere and the oceans, and it is this warming that prevents the planet from becoming a frozen wasteland. There are naturally occurring greenhouse gases—water vapor, carbon dioxide, ozone, methane and nitrous oxide—that create a benign greenhouse effect. Approximately one-third of the solar radiation that hits Earth is reflected back into space; some is absorbed by the atmosphere, but most is transferred to the land and oceans. However, human activities are causing greenhouse gas levels to increase as fossil fuels are burnt in ever-increasing amounts. At the same time, large tracts of rain forest that help remove carbon dioxide from the atmosphere are being cleared, intensifying the problem.

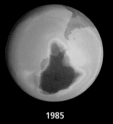 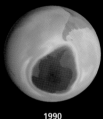 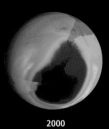

1985   1990   2000

← **Ozone filters out** harmful ultraviolet radiation. Since 1985, scientists have been measuring a widening hole in this atmospheric layer that opens up every year during the Antarctic summer.

← **Reef-building corals** require warm water to grow but excessive water temperatures, attributed to global warming, are thought to be the cause of coral bleaching. The corals do not die but become fragile and lose the algae that give them their distinct color.

→ **Current data indicates** the average global rate of sea level rise is 8 inches (20 cm) per century. This may not sound much, but for low-lying coral atolls, such as Arno Atoll in the Marshall Islands (*right*), this rise will be disastrous. Many atolls will be submerged by the end of the century.

↑ **Some of the predictions** of the effects of global warming suggest that sea levels will rise as the water locked up in the polar ice is released. However, there may be differences between the poles. In the Arctic, the Greenland ice sheet will melt at the margins but in Antarctica increased snowfall may offset melting.

GLOBAL TEMPERATURE CHANGES
The mean global surface temperature has increased by about 0.54 to 1.08°F (0.3 to 0.6°C) since the late nineteenth century, and by about 0.36 to 0.54°F (0.2 to 0.3°C) over the last 40 years. Recent years have been among the warmest since 1860. Warming is evident in both sea surface and land-based surface air temperatures. It should also be noted that the warming has not been globally uniform. The recent warming has been greatest between 40°N and 70°N latitude, although some areas, such as the North Atlantic Ocean, have cooled in the recent decades.

ANNUAL GLOBAL SURFACE MEAN TEMPERATURE ANOMALIES

°C | °F
Temperature
0.9 — 1.6 Land and Ocean
0.6 — 1.1
0.3 — 0.5
0.0 — 0.0
-0.3 — -0.5
-0.6 — -1.1

1880   1900   1920   1940   1960   1980   2000
Year

Conservation and restoration

With the increase in awareness of the environmental consequences of human activities, ways are being sought around the globe to protect Earth's precious resources and repair damage. The most obvious tool in protecting the oceans is effective conservation. This can take many forms and operate at different levels—from the local to the international, and from conserving a single population of one species to protecting a whole ecosystem. The international treaty that protects the Antarctic has proved that conservation can be successfully applied on a very large scale, and still allow research and ecotourism to flourish. Preventive measures such as strict fisheries controls, pollution controls and the creation of marine reserves are all being implemented. Where damage has already been done, such measures as restoration of habitats are being undertaken. For example, mangroves have been replanted on former shrimp-rearing ponds in southeast Asia and coastal defenses breached in Europe to recreate saltmarshes. Also, breeding programs have saved threatened species ranging from oysters to turtles.

↓ **Turtles have been bred** in large numbers to help restore local populations. Various schemes around the world, coupled with the protection of beaches where turtles come ashore to lay their eggs, will help to ensure that the seas continue to be populated by this ancient group of animals.

→ **Conservation cannot be** just simple prohibition of access. Ecotourism, such as non-invasive diving off Molokini Island, Hawaii (*pictured*), offers a way of providing an income that is not reliant on exploiting natural resources. It also allows the opportunity to show a wider audience the beauty of ocean life.

↗ **The Great Barrier Reef** has been protected as a World Heritage site since 1975. The reef park authority has to reconcile the different requirements of reef users, so that the reef can be enjoyed by all, without compromising its long-term viability.

→ **Artificial reefs increase** habitat diversity for fish and shellfish populations. They also help protect fish stocks and the seabed from trawling and dredging. Artificial reefs can be made from natural rocks, concrete blocks, old ships—or even aircraft.

↑ **The Galapagos Islands** have a unique position in the history of science; it was their unique animal communities that inspired Charles Darwin's studies and findings. The whole Galapagos archipelago is an Ecuadorian national park, home of the Charles Darwin Research Station and a World Heritage site.

ARTIFICIAL REEFS

The deliberate sinking of old ships to create artificial reefs is a well-known practice that has been carried out around the world. Artificial reefs enhance inshore fisheries, increase marine biodiversity and protect the seabed. In 2004, the sinking of the decommissioned frigate HMS *Scylla,* off the southwest coast of England, created an enormous artificial reef. This sunken vessel not only provides a new habitat for marine life, but is also used by scuba divers who enjoy safe diving on a wreck and avoid disturbing more sensitive sites. Considerable care was taken in preparing the vessel before sinking. All hazardous materials were removed and particular attention paid to cleaning off the toxic antifouling paint on the hull. Potential snagging hazards were removed, the doors were welded shut or open, and extra access holes for divers were opened up.

The future

Although only a small proportion of the world's population is directly involved in making a living from the oceans, everyone is affected by the state of the planet's largest surface component. The cradle of life on Earth, the oceans also control the weather, provide food and mineral wealth, and contain biological resources such as pharmaceutical products that have yet to be fully explored. Further, the oceans are a source of recreation. Today, a critical point has been reached, and the decisions made now about how oceans are used will have long-term consequences for the quality of life of future generations. Global warming will bring about changes in weather patterns and the ocean currents that moderate the climate extremes in many parts of the world. Rising sea levels caused by the melting of the polar ice will threaten coastal communities. However, effective controls on greenhouse gas emissions, fishing and waste dumping to ensure the health and survival of the Blue Planet are all achievable.

↓ **By monitoring changes** in individuals and populations of common animals, such as starfishes and sea stars (*pictured*), the health of the oceans can be measured. Early warning of problems can be given before irretrievable damage is done.

WHO OWNS THE OCEANS?

The oceans have always been regarded as a free and unrestricted resource. However, the need to reduce harmful human impact on the oceans requires a legal framework to establish ownership. Since 1960, international maritime law has given individual nations exclusive control over waters that extend for three nautical miles from their coastline. However, the nature of the seas requires a much broader-based approach. The United Nations Conference on Law of the Seas (UNCLOS) has sought since 1960 to make administration of the oceans, and their wealth of resources, a collective responsibility.

← **Turtles have been swimming** in the world's oceans for more than 200 million years. However, their numbers have fallen dramatically in the last 300 years. Some species are now extinct and others close to extinction. Human activities have had a major impact on marine ecosystems—through hunting or, indirectly, through habitat loss and pollution.

Factfile

Names of oceans and seas	Area square miles (km²)	Average depth feet (m)	Greatest known depth: feet (m)	Length miles (km)	Width miles (km)	Volume cubic miles (km³)
Planet Earth				(Equatorial circ.)*	(Polar circ.)*	
Planet Earth	196,930,000 (510,000,000)	–	–	24,902 (40,077)*	24,820 (39,942)*	–
The Continents (29.2% area of Earth)	57,510,000 (148,940,000)	–	–	–	–	–
The World Ocean (70.8% area of Earth)	139,420,000 (361,060,000)	12,430 (3790)	36,201 (11,034)	–	–	329,070,000 (1,370,740,000)
The three main ocean basins						
Pacific Ocean including marginal seas and Pacific section of the Southern Ocean	69,380,000 (179,680,000)	13,220 (4,030)	36,201 (11,034)	10,106 (16,264)	to 11,185 (to 18,000)	173,770,000 (723,840,000)
Atlantic Ocean including marginal seas (Black, Mediterranean, Caribbean etc), the Arctic Ocean and the Atlantic section of the Southern Ocean	41,110,000 (106,460,000)	10,920 (3330)	27,493 (8380)	13,360 (21,500)	to 4909 (7900)	85,200,000 (354,900,000)
Indian Ocean including marginal seas and the Indian Ocean section of the Southern Ocean	28,930,000 (74,920,000)	12,790 (3890)	24,460 (7,455)	6351 (10,220)	to 6338 (10,200)	70,100,000 (292,000,000)
The oceans						
Pacific Ocean, including marginal seas	65,590,000 (169,852,000)	13,127 (4001)	36,201 (11,034)	8637 (13,900)	to 11,185 (18,000)	163,100,000 (679,614,000)
Atlantic Ocean, including marginal seas	33,560,000 (86,915,000)	11,828 (3605)	28,233 (8605)	8774 (14,120)	to 4909 (7900)	75,200,000 (313,352,500)
Indian Ocean, including marginal seas	26,980,000 (69,876,000)	12,645 (3854)	24,460 (7455)	5841 (9400)	to 6338 (10,200)	64,720,000 (269,302,000)
Southern Ocean, including marginal seas	7,850,000 (20,327,000)	14,450 (4500)	23,736 (7235)	to 13,360 (21,500)	249–1678 (400–2700)	21,950,000 (91,471,500)
Arctic Ocean, including marginal seas	5,440,000 (14,090,000)	4690 (1430)	18,456 (5625)	to 3107 (5000)	to 1988 (3200)	4,100,000 (17,000,000)

THE OCEANS AND SUBDIVISIONS

IHO 23-4th: Limits of Oceans and Seas, Special Publication 23, 4th Edition June 2002, published by the International Hydrographic Bureau of the International Hydrographic Organization.

IHO	Names of oceans and seas	Area square miles (km²)	Average depth feet (m)	Greatest known depth: feet (m)	Length miles (km)	Width miles (km)	Volume cubic miles (km³)
	North Atlantic Ocean and subdivisions						
1	North Atlantic Ocean	–	–	28,233 (8605)	4636 (7460)	to 4909 (7900)	–
1.1	Skaggerak	12,970 (33,600)	1312 (400)	1969 (600)	150 (240)	80–90 (130–145)	3225 (13,440)
1.2	North Sea	222,100 (575,200)	308 (94)	2165 (660)	621 (1000)	93–373 (150–600)	12,974 (54,069)
1.3	Inner Seas (off West Coast Scotland)	4630 (12,000)	82 (25)	266 (81)	174 (280)	43–62 (70–100)	72 (300)
1.4	Irish Sea	40,000 (100,000)	125 (38)	576 (175)	130 (210)	150 (240)	912 (3800)
1.5	Bristol Channel	1160 (3000)	36 (11)	98 (30)	93 (150)	3–24 (5–40)	8 (33)
1.6	Celtic Sea	61,780 (160,000)	180 (55)	236 (72)	249 (400)	249 (400)	2112 (8800)
1.7	English Channel	34,700 (89,900)	272 (83)	394 (120)	249 (400)	21–112 (31–180)	1726 (7192)
1.7.1	Dover Strait	950 (2450)	46 (14)	180 (55)	43 (70)	18–25 (30–40)	8 (34)
1.8	Bay of Biscay	86,000 (223,000)	7874 (2400)	15,525 (4735)	317 (510)	342 (550)	128,419 (535,200)
1.9	Gulf of Guinea	324,360 (840,000)	12,796 (3900)	16,405 (5000)	435 (700)	746 (1200)	786,064 (3,276,000)
1.10	Caribbean Sea	1,049,500 (2,718,200)	8685 (2647)	25,218 (7686)	1678 (2700)	360–840 (600–1400)	1,726,430 (7,195,075)
1.11	Gulf of Mexico	615,000 (1,592,800)	4874 (1486)	12,425 (3787)	1100 (1770)	800 (1287)	567,929 (2,366,901)
1.12	Straits of Florida	48,270 (125,000)	1148 (350)	1641 (500)	311 (500)	155 (250)	10,498 (43,750)
1.13	Bay of Fundy	3600 (9300)	328 (100)	656 (200)	94 (151)	32 (52)	223 (930)
1.14	Gulf of St Lawrence	62,530 (162,000)	197 (60)	656 (200)	280 (450)	261 (420)	2332 (9720)
1.15	Labrador Sea	222,030 (575,000)	6562 (2000)	12,468 (3800)	to 870 (1400)	to 492 (820)	275,938 (1,150,000)

IHO	Names of oceans and seas	Area square miles (km²)	Average depth feet (m)	Greatest known depth: feet (m)
	Other subdivisions of the North Atlantic Ocean			
–	Bay of Campeche	65,640 (170,000)	1476 (450)	10,499 (3200)
–	Block Island Sound	190 (500)	59 (18)	66 (20)
–	Cabot Strait	1450 (3750)	1148 (350)	3281 (1000)
–	Canarias Sea	90,360 (234,000)	5906 (1800)	6562 (2000)
–	Cape Cod Bay	830 (2150)	82 (25)	115 (35)
–	Chaleur Bay	1680 (4350)	82 (25)	164 (50)
–	Chesapeake Bay	5800 (15,000)	23 (7)	80 (24)
–	Delaware Bay	970 (2500)	23 (7)	70 (21)
–	Denmark Strait	54,060 (140,000)	3281 (1000)	8203 (2500)
–	Gulf of Cadiz	14,480 (37,500)	1969 (600)	3281 (1000)
–	Gulf of Venezuela	10,190 (26,400)	8203 (2500)	9843 (3000)
–	Gulf of Honduras	30,890 (80,000)	2953 (900)	8203 (2500)
–	Gulf of Maine	90,700 (235,000)	492 (150)	1237 (377)
–	Long Bay	4630 (12,000)	49 (15)	131 (40)
–	Long Island Sound	1180 (3056)	82 (25)	328 (100)
–	Massachusetts Bay	820 (2123)	115 (35)	295 (90)
–	Mona Passage	8500 (22,000)	656 (200)	3281 (1000)
–	Nantucket Sound	1000 (2600)	59 (18)	131 (40)
–	North Channel	2320 (6000)	328 (100)	656 (200)
–	Onslow Bay	5790 (15,000)	66 (20)	164 (50)
–	Rhode Island Sound	580 (1500)	23 (7)	33 (10)
–	Saint George's Channel	4250 (11,000)	148 (45)	203 (62)
–	Sargasso Sea	1,448,020 (3,750,000)	16,405 (5000)	21,005 (6402)
–	Windward Passage	3480 (9000)	3281 (1000)	5545 (1690)
–	Yucatan Channel	16,600 (43,000)	3281 (1000)	6693 (2040)
	Baltic Sea and subdivisions (North Atlantic Ocean)			
2	Baltic Sea	163,000 (422,200)	180 (55)	1380 (421)
2.1	Central Baltic Sea	46,340 (120,000)	164 (50)	656 (200)
2.2	Gulf of Bothnia	45,200 (117,000)	200 (60)	965 (295)
2.2.1	Bothnian Sea	19,110 (49,500)	246 (75)	656 (200)
2.2.2	Bay of Bothnia	9270 (24,000)	246 (75)	656 (200)
2.3	Gulf of Finland	11,600 (30,000)	85 (26)	377 (115)
2.4	Sound Sea	25,100 (65,000)	82 (25)	164 (50)
2.5	Gulf of Riga	7000 (18,000)	131 (40)	144 (44)
2.6	The Sound	390 (1000)	131 (40)	164 (50)
2.7	The Great Belt	460 (1200)	85 (26)	164 (50)
2.8	The Little Belt	420 (1100)	85 (26)	164 (50)
2.9	Kattegat	9840 (25,485)	84 (26)	164 (50)
	Other subdivisions of the Baltic Sea (North Atlantic Ocean)			
–	Gulf of Gdansk	1660 (4296)	180 (55)	197 (60)
–	Kiel Bay	970 (2500)	131 (40)	131 (40)
–	Mecklenburger Bay	1000 (2600)	131 (40)	125 (38)
–	Pomeranian Bay	930 (2400)	66 (20)	164 (50)
	Mediterranean Sea and subdivisions (North Atlantic Ocean)			
3.1	Mediterranean Sea	969,000 (2,510,000)	4688 (1429)	16,897 (5150)

IHO	Names of oceans and seas	Area square miles (km²)	Average depth feet (m)	Greatest known depth: feet (m)	Length miles (km)	Width miles (km)	Volume cubic miles (km³)
Mediterranean Sea and subdivisions (North Atlantic Ocean) (continued)							
3.1.1	Mediterranean Sea, Western Basin	328,220 (850,000)	-	12,000 (3658)	1250 (2000)	497 (800)	-
3.1.1.1	Strait of Gibraltar	290 (750)	1200 (365)	3117 (950)	36 (58)	8 (13)	66 (274)
3.1.1.2	Alboran Sea	18,530 (48,000)	1969 (600)	3872 (1180)	249 (400)	48–120 (80–200)	6910 (28,800)
3.1.1.3	Balearic Sea	123,560 (320,000)	2461 (750)	4987 (1520)	497 (800)	120–420 (200–700)	57,587 (240,000)
3.1.1.4	Ligurian Sea	13,510 (35,000)	4265 (1300)	9300 (2850)	155 (250)	106 (170)	10,918 (45,500)
3.1.1.5	Tyrrhenian Sea	46,340 (120,000)	9515 (2900)	11,897 (3626)	475 (760)	60–300 (97–483)	83,501 (348,000)
3.1.2	Mediterranean Sea, Eastern Basin	640,990 (1,660,000)	-	16,897 (5150)	1250 (2000)	497 (800)	-
3.1.2.1	Adriatic Sea	52,220 (135,250)	1457 (444)	4035 (1324)	497 (800)	99 (160)	14,409 (60,051)
3.1.2.2	Strait of Sicily	19,310 (50,000)	328 (100)	1969 (600)	311 (500)	93 (150)	1200 (5000)
3.1.2.3	Ionian Sea	104,260 (270,000)	12,796 (3900)	16,000 (4900)	373 (600)	120–360 (200–600)	252,663 (1,053,000)
3.1.2.4	Aegean Sea	83,000 (214,000)	1969 (600)	11,627 (3543)	380 (611)	186 (399)	30,809 (128,400)
3.2	Sea of Marmara	4430 (11,474)	1620 (494)	4446 (1355)	175 (280)	50 (80)	1360 (5668)
3.3	Black Sea	196,000 (508,000)	4062 (1240)	7365 (2245)	730 (1175)	160 (260)	151,147 (629,920)
3.4	Sea of Azov	14,500 (37,555)	23 (7)	45 (13)	210 (340)	85 (135)	63 (263)
Other subdivisions of the Mediterranean Sea (North Atlantic Ocean)							
–	Dardanelles	100 (250)	180 (55)	300 (92)	38 (61)	0.75–4 (1.2–6.5)	3.3 (13.8)
–	Gulf of Lion	11,580 (30,000)	246 (75)	276 (84)	93 (150)	60–138 (100–230)	540 (2250)
–	Gulf of Venice	2010 (5200)	98 (30)	128 (39)	60 (95)	37 (60)	37 (156)
–	Bosporus	40 (105)	98 (30)	408 (124)	19 (30)	2.3 (3.7)	0.8 (3.2)
–	Sea of Crete	27,800 (72,000)	6562 (2000)	10,000 (3294)	249 (400)	130 (210)	34,552 (144,000)
–	Thracian Sea	3480 (9000)	246 (75)	328 (100)	93 (150)	50 (80)	162 (675)
Southern Atlantic Ocean and subdivisions							
4	South Atlantic Ocean	–	–	27,651 (8428)	4139 (6660)	to 4680 (to 7800)	–
4.1	River Plate	12,740 (33,000)	20 (6)	70 (21)	186 (300)	9 to 120 (15 to 200)	48 (198)
4.2	Scotia Sea	348,000 (900,000)	11,500 (4500)	27,651 (8428)	870 (1400)	497 (800)	755,830 (3,150,000)
4.3	Drake Passage	308,910 (800,000)	11,000 (3400)	15,600 (4800)	497 (800)	621 (1000)	652,653 (2,720,000)
Other subdivisions of the South Atlantic Ocean							
–	San Jorge Gulf	15,450 (40,000)	115 (35)	262 (80)	99 (160)	137 (220)	336 (1400)
–	San Matias Gulf	6950 (18,000)	98 (30)	230 (70)	99 (160)	24–90 (40–150)	130 (540)
–	Strait of Magellan	2240 (5800)	66 (20)	98 (30)	350 (560)	2–20 (3–32)	28 (116)
Indian Ocean and subdivisions							
5	Indian Ocean	26,980,000 (69,876,000)	12,645 (3854)	24,460 (7455)	5841 (9400)	to 6340 (10200)	64,720,000 (269,302,104)
5.1	Mozambique Channel	432,470 (1,120,000)	8531 (2600)	9843 (3000)	1000 (1600)	250–600 (400–950)	698,723 (2,912,000)
5.2	Gulf of Suez	4050 (10,500)	82 (25)	295 (90)	180 (290)	15–35 (24–56)	63 (263)
5.3	Gulf of Aqaba	1480 (3840)	2625 (800)	6070 (1850)	99 (160)	12–17 (19–27)	737 (3072)
5.4	Red Sea	169,100 (438,000)	1611 (491)	9974 (3040)	1200 (1930)	190 (305)	51,602 (215,058)
5.5	Gulf of Aden	205,000 (530,000)	4922 (1500)	17,586 (5360)	920 (1480)	300 (480)	190,757 (795,000)
5.6	Persian Gulf	90,000 (233,000)	115 (35)	328 (100)	615 (989)	35–210 (56–338)	1957 (8155)
5.7	Strait of Hormuz	4630 (12,000)	213 (65)	2953 (900)	99 (160)	35–60 (55–95)	187 (780)
5.8	Gulf of Oman	65,640 (170,000)	3937 (1200)	9843 (3000)	350 (560)	200 (320)	48,949 (204,000)
5.9	Arabian Sea	1,491,000 (3,862,000)	8970 (2734)	16,405 (5000)	to 1243 (2000)	to 1320 (2200)	2,533,521 (10,558,708)
5.10	Lakshadweep Sea	289,600 (750,000)	7874 (2400)	14,765 (4500)	to 932 (1500)	to 620 (1000)	431,903 (1,800,000)
5.11	Gulf of Mannar	11,740 (30,400)	3281 (1000)	5906 (1800)	99 (160)	80–170 (130–275)	7294 (30,400)
5.12	Palk Strait and Palk Bay	5780 (14,960)	98 (30)	295 (90)	85 (136)	40–85 (64–137)	108 (449)
5.13	Bay of Bengal	839,000 (2,173,000)	8500 (2600)	15,400 (4694)	1056 (1700)	994 (1600)	1,355,648 (5,649,800)
5.14	Andaman Sea	308,000 (797,700)	2854 (870)	12,392 (3777)	750 (1200)	400 (645)	166,522 (693,999)

IHO	Names of oceans and seas	Area sq. mi. (km²)	Average depth depth ft (m)	Greatest known depth ft (m)	Length miles (km)	Width miles (km)	Volume cubic miles (km³)
Indian Ocean and subdivisions (continued)							
5.15	Timor Sea	235,000 (615,000)	459 (140)	10,800 (3300)	609 (980)	435 (700)	20,659 (86,100)
5.15.1	Joseph Bonaparte Gulf	19,310 (50,000)	197 (60)	328 (100)	99 (160)	225 (360)	720 (3000)
5.16	Arafura Sea	250,990 (650,000)	230 (70)	12,000 (3660)	to 620 (1000)	to 435 (700)	10,918 (45,500)
5.16.1	Gulf of Carpentaria	120,000 (310,000)	164 (50)	230 (70)	544 (875)	120–390 (200–650)	3719 (15,500)
5.17	Great Australian Bight	366,830 (950,000)	7218 (2200)	14,765 (4500)	1740 (2800)	to 620 (1000)	501,487 (2,090,000)
Other subdivisions of the Indian Ocean							
–	Gulf of Bahrain	3280 (8500)	98 (30)	131 (40)	106 (170)	12–57 (20–95)	61 (255)
–	Strait of Tiran	440 (1150)	262 (80)	400 (122)	31 (50)	16 (25)	22 (92)
South China and Eastern Archipelagic Seas (Pacific Ocean)							
6	South China & Eastern Archipelagic Seas	–	–	–	–	–	–
6.1	South China Sea	895,400 (2,319,000)	5419 (1652)	16,456 (5016)	to 1182 (1970)	to 840 (1400)	919,231 (3,830,988)
6.2	Gulf of Tonkin	46,340 (120,000)	246 (75)	230 (70)	311 (500)	150 (250)	2160 (9000)
6.3	Gulf of Thailand	123,600 (320,000)	148 (45)	328 (100)	450 (725)	300–350 (500–560)	3455 (14,400)
6.4	Natuna Sea	135,150 (350,000)	148 (45)	328 (100)	559 (900)	60–390 (100–650)	3779 (15,750)
6.5	Malacca Strait	25,000 (65,000)	90 (27)	656 (200)	497 (800)	40–155 (65–249)	421 (1755)
6.6	Singapore Strait	1000 (2600)	131 (40)	164 (50)	65 (105)	10–18 (16–30)	25 (104)
6.7	Sunda Strait	4440 (11,500)	131 (40)	164 (50)	103 (165)	16–70 (26–110)	110 (460)
6.8	Java Sea	167,000 (433,000)	151 (46)	689 (210)	900 (1450)	261 (420)	4779 (19,918)
6.9	Makassar Strait	81,090 (210,000)	2625 (800)	6562 (2000)	497 (800)	80–230 (130–370)	40,311 (168,000)
6.10	Bali Sea	15,830 (41,000)	197 (60)	1805 (550)	311 (500)	75 (120)	590 (2460)
6.11	Flores Sea	93,000 (240,000)	6890 (2100)	16,860 (5140)	435 (700)	108–360 (180–600)	120,933 (504,000)
6.12	Sumba Strait	6490 (16,800)	328 (100)	2461 (750)	130 (210)	36–60 (60–100)	403 (1680)
6.13	Savu Sea	41,000 (105,000)	8859 (2700)	11,385 (3470)	404 (650)	155 (250)	68,025 (283,500)
6.14	Aru Sea	30,890 (80,000)	6562 (2000)	11,484 (3500)	311 (500)	78–102 (130–170)	38,391 (160,000)
6.15	Banda Sea	181,000 (470,000)	14,765 (4500)	24,409 (7440)	652 (1050)	228–330 (380–550)	507,486 (2,115,000)
6.16	Gulf of Bone	12,740 (33,000)	6562 (2000)	9843 (3000)	186 (300)	36–102 (60–170)	15,836 (66,000)
6.17	Ceram Sea	31,280 (81,000)	7218 (2200)	11,484 (3500)	360 (580)	68–120 (110–200)	42,758 (178,200)
6.18	Gulf of Berau	4250 (11,000)	230 (70)	328 (100)	137 (220)	12–42 (20–70)	185 (770)
6.19	Halmahera Sea	28,570 (74,000)	2461 (750)	3281 (1000)	186 (300)	186 (300)	13,317 (55,500)
6.20	Molucca Sea	77,000 (200,000)	9187 (2800)	15,780 (4810)	373 (600)	150–258 (250–430)	134,370 (560,000)
6.21	Gulf of Tomini	23,170 (60,000)	3937 (1200)	4922 (1500)	261 (420)	60–123 (100–205)	17,276 (72,000)
6.22	Celebes Sea	110,000 (280,000)	13,780 (4200)	20,406 (6220)	420 (675)	520 (837)	282,177 (1,176,000)
6.23	Sulu Sea	100,000 (260,000)	11,484 (3500)	18,400 (5600)	490 (790)	375 (603)	218,351 (910,000)
Other subdivisions of the South China and Eastern Archipelagic Seas							
–	Luzon Sea	81,090 (210,000)	2953 (900)	9843 (3000)	466 (750)	217 (350)	45,350 (189,000)
–	Luzon Strait	46,340 (120,000)	2953 (900)	6562 (2000)	217 (350)	249 (400)	25,914 (108,000)
–	Karimata Strait	52,130 (135,000)	492 (150)	656 (200)	280 (450)	120–228 (200–380)	4859 (20,250)
–	Yapen Strait	6180 (16,000)	3281 (1000)	3281 (1000)	155 (250)	40 (65)	3839 (16,000)
–	Cenderawasih Bay	18,150 (47,000)	2297 (700)	3281 (1000)	174 (280)	54–252 (90–420)	7894 (32,900)
North Pacific Ocean and subdivisions							
7	North Pacific Ocean	–	–	36,201 (11,034)	4499 (7240)	to 10,800 (18,000)	–
7.1	Philippine Sea	1,776,230 (4,600,000)	19,700 (6000)	34,578 (10,539)	to 1800 (3000)	to 1200 (2000)	6,622,513 (27,600,000)
7.2	Taiwan Strait	21,240 (55,000)	197 (60)	230 (70)	286 (460)	100–174 (160–280)	792 (3300)
7.3	East China Sea	284,000 (735,800)	574 (175)	8913 (2717)	684 (1100)	435 (700)	30,897 (128,765)
7.4	Yellow Sea	180,000 (466,200)	131 (40)	338 (103)	600 (960)	435 (700)	4475 (18,648)
7.4.1	Bo Hai	30,890 (80,000)	82 (25)	164 (50)	217 (350)	72–192 (120–320)	480 (2000)

IHO	Names of oceans and seas	Area square miles (km²)	Average depth feet (m)	Greatest known depth: feet (m)	Length miles (km)	Width miles (km)	Volume cubic miles (km³)
	North Pacific Ocean and subdivisions (continued)						
7.4.2	Liaodong Gulf	8110 (21,000)	82 (25)	164 (50)	112 (180)	75 (120)	126 (525)
7.5	Inland Sea of Japan	8500 (22,000)	121 (37)	197 (60)	233 (375)	6–105 (10–175)	195 (814)
7.6.1	Tatar Strait	23,940 (62,000)	246 (75)	3281 (1000)	311 (500)	78 (125)	1116 (4650)
7.7	Sea of Okhotsk	613,800 (1,589,700)	2749 (838)	12,001 (3658)	to 1020 (1700)	to 780 (1300)	319,649 (1,332,169)
7.8	Bering Sea	884,900 (2,291,900)	5075 (1547)	15,659 (4773)	1490 (2397)	990 (1593)	850,746 (3,545,569)
7.8.1	Gulf of Anadyr	37,070 (96,000)	246 (75)	492 (150)	200 (320)	250 (400)	1728 (7200)
7.9	Bering Strait	29,350 (76,000)	133 (40)	164 (50)	236 (380)	60–180 (100–300)	729 (3040)
7.10	Gulf of Alaska	592,000 (1,533,000)	8203 (2500)	16,405 (5000)	to 1200 (2000)	to 240 (400)	919,594 (3,832,500)
7.11	Coastal Waters of Southeast Alaska and British Colombia	13,900 (36,000)	9843 (3000)	12,468 (3800)	932 (1500)	to 99 (160)	25,914 (108,000)
7.12	Gulf of California	59,100 (153,000)	3937 (1200)	10,000 (3050)	750 (1200)	200 (320)	44,054 (183,600)
7.13	Gulf of Panama	8570 (22,200)	246 (75)	328 (100)	99 (160)	115 (185)	400 (1665)
	Other subdivisions of the North Pacific Ocean						
–	Amurskiy Liman	2160 (5600)	82 (25)	164 (50)	87 (140)	12–39 (20–65)	34 (140)
–	Bristol Bay	18,530 (48,000)	82 (25)	164 (50)	200 (320)	to 180 (300)	288 (1200)
–	Cheju Strait	3860 (10,000)	246 (75)	328 (100)	103 (165)	37 (60)	180 (750)
–	Gulf of Tehuantepec	12,160 (31,500)	984 (300)	3281 (1000)	326 (525)	75 (120)	2267 (9450)
–	Gulf of Santa Catalina	3670 (9500)	394 (120)	656 (200)	68 (110)	62 (100)	274 (1140)
–	Hecate Strait	7920 (20,500)	1476 (450)	3281 (1000)	160 (257)	40–60 (64–129)	2214 (9225)
–	Strait of La Perouse	4830 (12,500)	246 (75)	328 (100)	130 (210)	27–51 (45–85)	225 (938)
–	Korea Bay	13,900 (36,000)	98 (30)	164 (50)	165 (265)	121 (195)	259 (1080)
–	Korea Strait	2510 (6500)	295 (90)	328 (100)	62 (100)	30–48 (50–80)	140 (585)
–	Queen Charlotte Sound	7410 (19,200)	1575 (480)	3281 (1000)	124 (200)	75 (120)	2211 (9216)
–	Sakhalin Gulf	2900 (7500)	98 (30)	164 (50)	47 (75)	36–96 (60–160)	54 (225)
–	Santa Barbara Channel	2320 (6000)	1312 (400)	1641 (500)	75 (120)	31 (50)	576 (2400)
–	San Pedro Channel	770 (2000)	394 (120)	656 (200)	28 (45)	28 (45)	58 (240)
–	Sea of Japan	377,600 (978,000)	5748 (1752)	12,276 (3742)	1740 (2800)	to 540 (900)	411,137 (1,713,456)
–	Strait of Georgia	1450 (3750)	328 (100)	1200 (370)	138 (222)	17 (28)	90 (375)
–	Strait of Juan de Fuca	1310 (3400)	394 (120)	900 (275)	80–100 (130–160)	16 (25)	98 (408)
–	Gulf of Shelikhova	69,500 (180,000)	410 (125)	1624 (495)	420 (670)	185 (300)	5399 (22,500)
	South Pacific Ocean and subdivisions						
8	South Pacific Ocean	–	–	35,704 (10,882)	4139 (6660)	to 10,800 (18,000)	–
8.1	Bismarck Sea	194,227 (503,000)	6600 (2000)	8200 (2500)	497 (800)	249 (400)	241,386 (1,006,000)
8.2	Solomon Sea	278,019 (720,000)	14,765 (4500)	29,988 (9140)	621 (1000)	497 (800)	777,425 (3,240,000)
8.3	Coral Sea	1,849,000 (4,790,000)	7870 (2400)	25,134 (7661)	1400 (2250)	1500 (2414)	2,758,421 (11,496,000)
8.3.1	Torres Strait	10,430 (27,000)	246 (75)	328 (100)	130 (210)	80 (130)	486 (2025)
8.3.2	Great Barrier Reef (Coastal Waters)	96,530 (250,000)	197 (60)	328 (100)	1087 (1750)	30–180 (50–300)	3599 (15,000)
8.3.3	Gulf of Papua	14,670 (38,000)	213 (65)	328 (100)	95 (150)	225 (360)	593 (2470)
8.4	Tasman Sea	1,545,000 (4,000,000)	9023 (2750)	17,000 (5200)	1243 (2000)	1400 (2250)	2,639,407 (11,000,000)
8.4.1	Bass Strait	28,950 (75,000)	210 (60)	262 (80)	224 (360)	150 (240)	1080 (4500)
	Other subdivisions of the South Pacific Ocean						
–	Bay of Plenty	3090 (8000)	328 (100)	820 (250)	99 (160)	37 (60)	192 (800)
–	Cook Strait	2120 (5500)	420 (128)	3445 (1050)	81 (130)	14 (23)	169 (704)
–	Foveaux Strait	970 (2500)	98 (30)	164 (50)	43 (70)	22 (35)	18 (75)
–	Gulf of Guayaquil	5410 (14,000)	82 (25)	197 (60)	124 (200)	87 (140)	84 (350)
–	Hawke Bay	540 (1400)	115 (35)	197 (60)	50 (80)	28 (45)	12 (49)

IHO	Names of oceans and seas	Area square miles (km²)	Average depth feet (m)	Greatest known depth: feet (m)	Length miles (km)	Width miles (km)	Volume cubic miles (km³)
	Arctic Ocean and subdivisions						
9	Arctic Ocean	5,440,000 (14,090,000)	4690 (1430)	18,456 (5625)	to 3000 (5000)	to 1920 (3200)	4,834,603 (20,148,700)
9.1	East Siberian Sea	361,000 (936,000)	328 (100)	510 (155)	777 (1250)	497 (800)	22,459 (93,600)
9.2	Laptev Sea	250,900 (649,800)	1896 (578)	9774 (2980)	528 (850)	497 (800)	90,120 (375,584)
9.3	Kara Sea	340,000 (880,000)	417 (127)	2034 (620)	932 (1500)	559 (900)	26,816 (111,760)
9.4	Barents Sea	542,000 (1,405,000)	750 (229)	1969 (600)	808 (1300)	650 (1050)	77,201 (321,745)
9.5	White Sea	36,680 (95,000)	200 (60)	1115 (340)	261 (420)	249 (400)	1368 (5700)
9.6	Greenland Sea	353,320 (915,000)	4750 (1450)	16,000 (4800)	808 (1300)	621 (1000)	318,349 (1,326,750)
9.7	Norwegian Sea	328,220 (850,000)	5254 (1600)	13,020 (3970)	870 (1400)	684 (1100)	326,327 (1,360,000)
9.8	Iceland Sea	111,980 (290,000)	3701 (1128)	9843 (3000)	404 (650)	280 (450)	78,491 (327,120)
9.9	Davis Strait	115,840 (300,000)	1476 (450)	4922 (1500)	404 (650)	200–400 (322–644)	32,393 (135,000)
9.10	Hudson Strait	37,070 (96,000)	1641 (500)	3090 (942)	497 (800)	40–150 (65–240)	11,517 (48,000)
9.11	Hudson Bay	475,800 (1,232,300)	420 (128)	600 (183)	590 (950)	590 (950)	37,848 (157,734)
9.12	Baffin Bay	266,000 (689,000)	6234 (1900)	7000 (2100)	900 (1450)	68–400 (110–650)	314,113 (1,309,100)
9.13	Lincoln Sea	57,920 (150,000)	2461 (750)	9394 (2863)	249 (400)	236 (380)	26,994 (112,500)
9.14	Northwestern Passages	–	492 (150)	656 (200)	–	–	–
9.15	Beaufort Sea	184,000 (476,000)	3239 (1004)	15,360 (4682)	684 (1100)	404 (650)	114,671 (477,904)
9.16	Chukchi Sea	225,000 (582,000)	253 (77)	7218 (2200)	559 (900)	435 (700)	10,753 (44,814)
	Other subdivisions of the Arctic Ocean						
–	Fox Basin	55,600 (144,000)	492 (150)	656 (200)	280 (450)	249 (400)	5183 (21,600)
–	James Bay	30,890 (80,000)	164 (50)	230 (70)	275 (443)	135 (217)	960 (4000)
–	Kane Basin	7720 (20,000)	492 (150)	656 (200)	124 (200)	62 (100)	720 (3000)
–	Pechora Sea	34,750 (90,000)	20 (6)	689 (210)	249 (400)	162 (260)	130 (540)
–	Wandel Sea	28,960 (75,000)	1148 (350)	1955 (596)	186 (300)	155 (250)	6299 (26250)
	Southern Ocean and subdivisions						
10	Southern Ocean	7,850,000 (20,327,000)	14,750 (4500)	23,736 (7235)	to 12,900 (21,500)	240–1620 (400–2700)	21,948,232 (91,471,500)
10.1	Weddell Sea	1,080,000 (2,800,000)	13,124 (4000)	16,405 (5000)	to 1200 (to 2000)	to 1200 (2000)	2,687,397 (11,200,000)
10.2	Lazarev Sea	185,350 (480,000)	11,484 (3500)	13,124 (4000)	497 (800)	373 (600)	403,109 (1,680,000)
10.3	Riiser-Larsen Sea	260,640 (675,000)	12,468 (3800)	13,124 (4000)	559 (900)	404 (650)	615,462 (2,565,000)
10.4	Cosmonauts Sea	386,140 (1,000,000)	14,108 (4300)	16,405 (5000)	621 (1000)	621 (1000)	1,031,768 (4,300,000)
10.5	Cooperation Sea	386,140 (1,000,000)	9515 (2900)	13,124 (4000)	621 (1000)	621 (1000)	695,844 (2,900,000)
10.6	Davis Sea	347,520 (900,000)	6562 (2000)	9843 (3000)	621 (1000)	559 (900)	431,903 (1,800,000)
10.6.1	Tryoshnikova Gulf	34,750 (90,000)	6562 (2000)	9843 (3000)	280 (450)	to 240 (400)	43,190 (180,000)
10.7	Mawson Sea	96,530 (250,000)	6562 (2000)	9843 (3000)	311 (500)	311 (500)	119,973 (500,000)
10.8	Dumont d'Urville Sea	185,350 (480,000)	5906 (1800)	13,124 (4000)	497 (800)	373 (600)	207,313 (864,000)
10.9	Somov Sea	100,400 (260,000)	6562 (2000)	6562 (2000)	404 (650)	249 (400)	124,772 (520,000)
10.10	Ross Sea	370,000 (960,000)	656 (200)	2625 (800)	684 (1100)	621 (1000)	46,070 (192,000)
10.10.1	McMurdo Sound	3860 (10,000)	3281 (1000)	3281 (1000)	92 (148)	46 (74)	2399 (10,000)
10.11	Amundsen Sea	297,330 (770,000)	6562 (2000)	9843 (3000)	684 (1100)	435 (700)	369,517 (1,540,000)
10.12	Bellingshausen Sea	173,760 (450,000)	6562 (2000)	9843 (3000)	435 (700)	404 (650)	215,952 (900,000)
10.13	Drake Passage	308,910 (800,000)	11,000 (3400)	15,600 (4800)	497 (800)	621 (1000)	652,653 (2,720,000)
10.14	Bransfield Strait	26,260 (68,000)	1148 (350)	1641 (500)	249 (400)	106 (170)	5711 (23,800)
	Inland Seas (salt lakes not seas)						
–	Caspian Sea	152,239 (394,299)	591 (180)	3104 (946)	746 (1200)	102–270 (170–450)	17,030 (70,974)
–	Aral Sea	13,000 (33,800)	52 (16)	223 (68)	266 (428)	176 (284)	130 (541)
–	Dead Sea	394 (1020)	313 (96)	1310 (399)	48 (78)	10 (15)	23 (97)
–	Salton Sea	344 (890)	30 (9)	51 (16)	30 (48)	10 (16)	2 (8)
–	Sea of Galilee	64 (166)	79 (24)	157 (48)	13 (21)	7 (11)	1 (4)

IHO	Names of oceans and seas	Area square miles (km²)	Average depth feet (m)	Greatest known depth: feet (m)	Length miles (km)	Width miles (km)	Volume cubic miles (km³)
	Largest subdivisions of the oceans by surface area						
8.3	Coral Sea	1,849,000 (4,790,000)	7870 (2400)	25,134 (7661)	1400 (2250)	1500 (2414)	2,758,421 (11,496,000)
7.1	Philippine Sea	1,776,230 (4,600,000)	19,700 (6000)	34,578 (10,539)	to 1800 (3000)	to 1200 (2000)	6,622,513 (27,600,000)
8.4	Tasman Sea	1,545,000 (4,000,000)	9023 (2750)	17,000 (5200)	1243 (2000)	1400 (2250)	2,639,407 (11,000,000)
5.9	Arabian Sea	1,491,000 (3,862,000)	8970 (2734)	16,405 (5000)	to 1243 (2000)	to 1320 (2200)	2,533,521 (10,558,708)
–	Sargasso Sea	1,448,020 (3,750,000)	16,405 (5000)	21,005 (6402)	1864 (3000)	994 (1600)	4,498,990 (18,750,000)
10.1	Weddell Sea	1,080,000 (2,800,000)	13,124 (4000)	16,405 (5000)	to 1200 (2000)	to 1200 (2000)	2,687,397 (11,200,000)
1.10	Caribbean Sea	1,049,500 (2,718,200)	8685 (2647)	25,218 (7686)	1678 (2700)	360–840 (600–1400)	1,726,430 (7,195,075)
3.1	Mediterranean Sea	969,000 (2,510,000)	4688 (1429)	16,897 (5150)	2500 (4000)	500 (800)	860,636 (3,586,790)
6.1	South China Sea	895,400 (2,319,000)	5419 (1652)	16,456 (5016)	to 1182 (1970)	to 840 (1400)	919,231 (3,830,988)
7.8	Bering Sea	884,900 (2,291,900)	5075 (1547)	15,659 (4773)	1490 (2397)	990 (1593)	850,746 (3,545,569)
5.13	Bay of Bengal	839,000 (2,173,000)	8500 (2600)	15,400 (4694)	1056 (1700)	994 (1600)	1,355,648 (5,649,800)
1.11	Gulf of Mexico	615,000 (1,592,800)	4874 (1486)	12,425 (3787)	1100 (1770)	800 (1287)	567,929 (2,366,901)
7.7	Sea of Okhotsk	613,800 (1,589,700)	2749 (838)	12,001 (3658)	to 1020 (1700)	to 780 (1300)	319,649 (1,332,169)
7.10	Gulf of Alaska	592,000 (1,533,000)	8203 (2500)	16,405 (5000)	to 1200 (2000)	to 240 (400)	919, 594 (3,832,500)
9.4	Barents Sea	542,000 (1,405,000)	750 (229)	2000 (600)	800 (1300)	650 (1050)	77,201 (321,745)
9.11	Hudson Bay	475,800 (1,232,300)	420 (128)	600 (183)	590 (950)	590 (950)	37,848 (157,734)
5.1	Mozambique Channel	432,470 (1,120,000)	8531 (2600)	10,000 (3000)	1000 (1600)	250–600 (400–950)	698,723 (2,912,000)
10.4	Cosmonauts Sea	386,140 (1,000,000)	14,108 (4300)	16,405 (5000)	621 (1000)	621 (1000)	1,031,768 (4,300,000)
10.5	Cooperation Sea	386,140 (1,000,000)	9515 (2900)	13,124 (4000)	621 (1000)	621 (1000)	695,844 (2,900,000)
–	Sea of Japan	377,600 (978,000)	5748 (1752)	12,276 (3742)	1740 (2800)	to 540 (900)	411,137 (1,713,456)
10.10	Ross Sea	370,000 (960,000)	656 (200)	2625 (800)	684 (1100)	621 (1000)	46,070 (192,000)
5.17	Great Australian Bight	366,830 (950,000)	7218 (2200)	14,765 (4500)	1740 (2800)	to 620 (1000)	501,487 (2,090,000)
9.1	East Siberian Sea	361,000 (936,000)	328 (100)	510 (155)	777 (1250)	497 (800)	22,459 (93,600)
9.6	Greenland Sea	353,320 (915,000)	4750 (1450)	16,000 (4800)	808 (1300)	621 (1000)	318,349 (1,326,750)
4.2	Scotia Sea	348,000 (900,000)	11,500 (3500)	27,651 (8428)	870 (1400)	497 (800)	755,830 (3,150,000)
10.6	Davis Sea	347,520 (900,000)	6562 (2000)	9843 (3000)	621 (1000)	559 (900)	431,903 (1,800,000)
9.3	Kara Sea	340,000 (880,000)	417 (127)	2034 (620)	932 (1500)	559 (900)	26,816 (111,760)
9.7	Norwegian Sea	328,220 (850,000)	5254 (1600)	13,020 (3970)	870 (1400)	684 (1100)	326,327 (1,360,000)
1.9	Gulf of Guinea	324,360 (840,000)	12,796 (3900)	16,405 (5000)	435 (700)	746 (1200)	786,064 (3,276,000)
10.13 + 4.3	Drake Passage	308,910 (800,000)	11,000 (3400)	15,600 (4800)	497 (800)	621 (1000)	652,653 (2,720,000)
5.14	Andaman Sea	308,000 (797,700)	2854 (870)	12,392 (3777)	750 (1200)	400 (645)	166,522 (693,999)
10.11	Amundsen Sea	297,330 (770,000)	6562 (2000)	9843 (3000)	684 (1100)	435 (700)	369,517 (1,540,000)
5.10	Lakshadweep Sea	289,600 (750,000)	7874 (2400)	14,765 (4500)	to 932 (1500)	to 620 (1000)	431,903 (1,800,000)
7.3	East China Sea	284,000 (735,800)	574 (175)	8913 (2717)	684 (1100)	435 (700)	30,897 (128,765)
8.2	Solomon Sea	278,019 (720,000)	14,765 (4500)	29,988 (9140)	621 (1000)	497 (800)	777,425 (3,240,000)
9.12	Baffin Bay	266,000 (689,000)	6234 (1900)	7000 (2100)	900 (1450)	70–400 (110–650)	314,113 (1,309,100)
10.3	Riiser-Larsen Sea	260,640 (675,000)	12,468 (3800)	13,124 (4000)	559 (900)	404 (650)	615,462 (2,565,000)
5.16	Arafura Sea	250,990 (650,000)	230 (70)	12,000 (3660)	to 620 (1000)	to 435 (700)	10,918 (45,500)
9.2	Laptev Sea	250,900 (649,800)	1896 (578)	9774 (2980)	528 (850)	497 (800)	90,120 (375,584)
5.15	Timor Sea	235,000 (615,000)	459 (140)	10,800 (3300)	609 (980)	435 (700)	20,659 (86,100)
9.16	Chukchi Sea	225,000 (582,000)	253 (77)	7218 (2200)	559 (900)	435 (700)	10,753 (44,814)
1.2	North Sea	222,100 (575,200)	308 (94)	2165 (660)	621 (1000)	93–373 (150–600)	12,974 (54,069)
1.15	Labrador Sea	222,030 (575,000)	6562 (2000)	12,468 (3800)	to 870 (1400)	to 492 (820)	275,938 (1,150,000)
5.5	Gulf of Aden	205,000 (530,000)	4922 (1500)	17,586 (5360)	920 (1480)	300 (480)	190,757 (795,000)
3.3	Black Sea	196,000 (508,000)	4062 (1240)	7365 (2245)	730 (1175)	160 (260)	151,147 (629,920)
8.1	Bismarck Sea	194,227 (503,000)	6600 (2000)	8200 (2500)	497 (800)	249 (400)	241,386 (1,006,000)
10.2	Lazarev Sea	185,350 (480,000)	11,484 (3500)	13,124 (4000)	497 (800)	373 (600)	403,109 (1,680,000)
10.8	Dumont d'Urville Sea	185,350 (480,000)	5906 (1800)	13,124 (4000)	497 (800)	373 (600)	207,313 (864,000)
9.15	Beaufort Sea	184,000 (476,000)	3239 (1004)	15,360 (4682)	684 (1100)	404 (650)	114,671 (477,904)

IHO	Names of oceans and seas	Area square miles (km²)	Average depth feet (m)	Greatest known depth: feet (m)	Length miles (km)	Width miles (km)	Volume cubic miles (km³)
	Largest subdivisions of the oceans by surface area (continued)						
6.15	Banda Sea	181,000 (470,000)	14,765 (4500)	24,409 (7440)	652 (1050)	228–330 (380–550)	507,486 (2,115,000)
7.4	Yellow Sea	180,000 (466,200)	131 (40)	338 (103)	600 (960)	435 (700)	4475 (18,648)
10.12	Bellingshausen Sea	173,760 (450,000)	6562 (2000)	9843 (3000)	435 (700)	404 (650)	215,952 (900,000)
5.4	Red Sea	169,100 (438,000)	1611 (491)	9974 (3040)	1200 (1930)	190 (305)	51,602 (215,058)
6.8	Java Sea	167,000 (433,000)	151 (46)	689 (210)	900 (1450)	260 (420)	4779 (19,918)
2	Baltic Sea	163,000 (422,200)	180 (55)	1380 (421)	795 (1280)	24–324 (40–540)	5572 (23,221)

MARINE PARKS AND RESERVES

The International Union for Conservation of Nature and Natural Resources (IUCN) management categories

Category Ia: Strict nature reserve: Protected area managed mainly for science

Category Ib: Wilderness area: Protected area managed mainly for wilderness protection

Category II: National park: Protected area managed mainly for ecosystem protection and recreation

Category III: Natural monument: Protected area managed mainly for conservation of specific natural features

Category IV: Habitat/species management area: Protected area managed mainly for conservation through management intervention

Category V: Protected landscape/seascape: Protected area managed mainly for landscape/seascape conservation and recreation (not included in factfile)

Category VI: Managed resource protected area: Protected area managed mainly for the sustainable use of natural ecosystems (not included in factfile)

NOTES

LONGITUDE AND LATITUDE

Positive longitude numbers = east
Negative longitude numbers = west
Positive latitude numbers = north
Negative latitude numbers = south

Area name	Designate	Country	Longitude	Latitude
International Union for Conservation of Nature and Natural Resources (IUCN) category Ia: Strict nature reserves				
Cape Rodney—Okakari Point	Marine Reserve	New Zealand	174.79475	-36.26488
Cartier Island	Marine Reserve	Australia	123.55522	-12.53078
Cayos Miskitos	Marine Reserve	Nicaragua	-82.81026	14.39545
Dwesa-Cwebe	Marine Nature Reserve	South Africa	28.87245	-32.26708
Elizabeth and Middleton Reefs	Marine National Nature Reserve	Australia	159.075	-29.7
Heard Island and McDonald Islands	Marine Reserve	Heard and McDonald Islands	73.34232	-52.34843
Hluleka	Marine Sanctuary	South Africa	29.32397	-31.81069
Kapiti	Marine Reserve	New Zealand	174.94992	-40.85648
Kermadec Islands	Marine Reserve	New Zealand	178.5	-30.333
Long Bay—Okura	Marine Reserve	New Zealand	174.74723	-36.67141
Long Island-Kokomohua	Marine Reserve	New Zealand	174.28	-41.12
Mayor Island (Tuhua)	Marine Reserve	New Zealand	176.25	-37.27
Mermaid Reef	Marine National Nature Reserve	Australia	119.625	-17.09167
Mkambati	Marine Sanctuary	South Africa	30.00114	-31.28687
Motu Manawa—Pollen Island	Marine Reserve	New Zealand	174.67894	-36.86813
Piopiotahi (Milford Sound)	Marine Reserve	New Zealand	167.89318	-44.63278
Pohatu	Marine Reserve	New Zealand	173.02197	-43.87811

Area name	Designate	Country	Longitude	Latitude
International Union for Conservation of Nature and Natural Resources (IUCN) category Ia: Strict nature reserves (continued)				
Poor Knights Islands	Marine Reserve	New Zealand	174.73	-35.471
Shark Bay	Marine Park	Australia	113.75	-25.8
Tasmanian Seamounts	Marine Reserve	Australia	147.31226	-44.39363
Te Angiangi	Marine Reserve	New Zealand	176.85047	-40.1587
Te Awaatu Channel (The Gut)	Marine Reserve	New Zealand	166.94956	-45.29701
Te Tapuwae O Rongokako	Marine Reserve	New Zealand	178.204	-38.606
Tonga Island	Marine Reserve	New Zealand	173.07021	-40.88108
Westhaven—Te Taitapu	Marine Reserve	New Zealand	172.53259	-40.61738
International Union for Conservation of Nature and Natural Resources (IUCN) category Ib: Wilderness area				
Lampi	Marine National Park	Myanmar	98.10634	10.63511
International Union for Conservation of Nature and Natural Resources (IUCN) category II: National park				
Abrolhos	Marine National Park	Brazil	-38.66116	-17.92351
Ambodilaitry Masoala	Marine Park	Madagascar	50.17556	-15.99678
Aragusuku-jima Maibushi	Marine Park	Japan	124	24.5
Awa-oshima	Marine Park	Japan	134.6667	33.66667
Baie Ternaie	Marine National Park	Seychelles	55.37426	-4.63846
Bloody Bay—Jackson Point	Marine Park	Cayman Islands	-80.0745	19.68889
Bowse Bluff—Rum Point (Grand Cayman)	Marine Park	Cayman Islands	-81.26767	19.37249
Chumbe Island Coral Park (CHICOP)	Marine Sanctuary	Tanzania, United Republic of	39.17486	-6.27774
Curieuse	Marine National Park	Seychelles	55.72895	-4.29157
Diamond Reef	Marine Park	Antigua and Barbuda	-61.86178	17.19565
Dick Sessingers Bay—Beach Point	Marine Park	Cayman Islands	-79.86315	19.68453
Fernando de Noronha	Marine National Park	Brazil	-32.40322	-3.86214
Folkestone	Marine Reserve	Barbados	-59.64	13.13
Fukue	Marine Park	Japan	129	32.5
Garig Gunak Barlu National Park	Marine Park	Australia	132.18	-11.34
Genkai	Marine Park	Japan	130	33.5
Goshikigahama	Marine Park	Japan	135	35.66667
Great Australian Bight	Marine Park	Australia	131.13495	-31.56669
Gulf of Kutch	Marine National Park	India	69.61281	22.45967
Gulf of Mannar	Marine National Park	India	78.72723	9.10104
Hamasaka	Marine Park	Japan	134.5	35.5
Hol Chan	Marine Reserve	Belize	-87.99946	17.87116
Ifaho	Marine Park	Madagascar	50.35025	-15.78074
Iki Tatsunoshima	Marine Park	Japan	129.5	33.5
Iki Tsumagashima	Marine Park	Japan	129.5	33.5
Isla Exposición	Marine National Park	Honduras	-97.68306	13.33306
Islas del Cisne	Marine National Park	Honduras	-84	17.42303
Jennifer Bay—Deep Well (Cayman Brac)	Marine Park	Cayman Islands	-79.78736	19.71108
Kasari Hanto Higashi Kaigan	Marine Park	Japan	129.5	28.5
Kashinishi	Marine Park	Japan	133.5	33
Katsuura	Marine Park	Japan	140.3333	35.16667
Kisite	Marine National Park	Kenya	39.36247	-4.714
Malindi	Marine National Park	Kenya	40.14334	-3.25561
Mananara-Nord	Marine Park	Madagascar	49.84698	-16.3243
Maziwi Island	Marine Reserve	Tanzania, United Republic of	39.0763	-5.50171
Moheli	Marine Park	Comoros	43.76667	-12.16667
Mombasa	Marine National Park	Kenya	39.76151	-3.98634
Montego Bay	Marine Park	Jamaica	-77.95751	18.47488

Ningaloo	Marine Park	Australia	113.83153	-22.53249
Ningaloo (Commonwealth Waters)	Marine Park	Australia	113.63858	-22.89348
North West Point—West Bay Cemetery	Marine Park	Cayman Islands	-81.40827	19.3672
Nukunonu Marine Conservation Area	Marine Reserve	Tokelau	-171.81666	-9.11667
Old Pageant Beach—Sand Cay Apartments	Marine Park	Cayman Islands	-81.3897	19.29272
Palaster Reef	Marine National Park	Antigua and Barbuda	-61.74716	17.51948
Port Launay	Marine National Park	Seychelles	55.39013	-4.65148
Preston Bay—Main Channel MP	Marine Park	Cayman Islands	-80.08978	19.65594
Pulau Aur	Marine Park	Malaysia	104.52	2.42667
Pulau Besar	Marine Park	Malaysia	103.97833	2.44167
Pulau Chebeh	Marine Park	Malaysia	104.10833	2.93333
Pulau Ekor Tebu	Marine Park	Malaysia	103.03111	5.73194
Pulau Goal	Marine Park	Malaysia	103.97	2.53667
Pulau Harimau	Marine Park	Malaysia	103.94333	2.55833
Pulau Hujung	Marine Park	Malaysia	103.95333	2.49167
Pulau Jahat	Marine Park	Malaysia	104.178	2.70522
Pulau Kaca	Marine Park	Malaysia	100.04083	6.07278
Pulau Kapas	Marine Park	Malaysia	103.26667	5.22083
Pulau Kuraman	Marine Park	Malaysia	115.12917	5.22333
Pulau Labas	Marine Park	Malaysia	104.088	2.87752
Pulau Lang Tengah	Marine Park	Malaysia	102.9	5.67972
Pulau Lembu	Marine Park	Malaysia	100.05806	6.07278
Pulau Lima	Marine Park	Malaysia	103.06167	5.75667
Pulau Mensirip	Marine Park	Malaysia	103.95167	2.55111
Pulau Mentinggi	Marine Park	Malaysia	104.13833	2.31444
Pulau Nyireh	Marine Park	Malaysia	103.66667	4.84472
Pulau Payar	Marine Park	Malaysia	100.04083	6.06333
Pulau Pemanggil	Marine Park	Malaysia	104.32667	2.57389
Pulau Perhentian Besar	Marine Park	Malaysia	102.75833	5.9
Pulau Perhentian Kecil	Marine Park	Malaysia	102.725	5.91
Pulau Pinang	Marine Park	Malaysia	103.00167	5.74083
Pulau Rawa	Marine Park	Malaysia	103.9775	2.52
Pulau Redang	Marine Park	Malaysia	103.01183	5.7846
Pulau Rusukan Besar	Marine Park	Malaysia	115.13667	5.18833
Pulau Rusukan Kecil	Marine Park	Malaysia	115.14333	5.20278
Pulau Segantang	Marine Park	Malaysia	99.92694	6.04333
Pulau Sembilang	Marine Park	Malaysia	103.89167	2.68778
Pulau Sepoi	Marine Park	Malaysia	104.064	2.90319
Pulau Sibu	Marine Park	Malaysia	104.0725	2.21417
Pulau Sibu Hujung	Marine Park	Malaysia	104.06667	2.23333
Pulau Sri Buat	Marine Park	Malaysia	103.91167	2.69472
Pulau Susu Dara	Marine Park	Malaysia	102.675	5.96083
Pulau Tengah	Marine Park	Malaysia	103.9625	2.47694
Pulau Tenggol	Marine Park	Malaysia	103.66667	4.80667
Pulau Tinggi	Marine Park	Malaysia	104.16667	2.25
Pulau Tioman	Marine Park	Malaysia	104.16667	2.76667
Pulau Tokong Bahara	Marine Park	Malaysia	104.128	2.71261
Pulau Tulai	Marine Park	Malaysia	104.10056	2.90278
Sanganeb Atoll	Marine National Park	Sudan	37.417	19.75
Scotts Anchorage—White Bay (Cayman Brac)	Marine Park	Cayman Islands	-79.87941	19.69695

Area name	Designate	Country	Longitude	Latitude
International Union for Conservation of Nature and Natural Resources (IUCN) category II: National park (continued)				
Silhouette Marine	Marine National Park	Seychelles	55.23057	-4.48783
Spanish Cove Resort—Jetty (Grand Cayman)	Marine Park	Cayman Islands	-81.39604	19.3951
Ste. Anne	Marine National Park	Seychelles	55.50125	-4.62099
Stellwagen Banks	Marine Sanctuary	United States	-70.317	41.417
Tampolo	Marine Park	Madagascar	49.96889	-15.76052
Victoria House—Treasure Island Resort	Marine Park	Cayman Islands	-81.38926	19.33573
Wandur	Marine National Park	India	92.65685	11.57063
Watamu	Marine National Park	Kenya	40.07272	-3.33414
Wilsons Promontory	Marine Reserve	Australia	146.243	-39.08889
International Union for Conservation of Nature and Natural Resources (IUCN) category III: Natural monument				
Wreck of the Rhone	Marine Park	Virgin Islands (British)	-64.55311	18.36863
International Union for Conservation of Nature and Natural Resources (IUCN) category IV: Habitat/species management areas				
Anse Chastanet Reefs	Marine Reserve	Saint Lucia	-61.083	13.867
Anse Cochon artificial reef	Marine Reserve	Saint Lucia	-61.05	13.92597
Anse Galet-Anse Cochon reefs	Marine Reserve	Saint Lucia	-61.05	13.92597
Anse L'Ivrogne Reef	Marine Reserve	Saint Lucia	-61.067	13.8
Anse Mamin Reef	Marine Reserve	Saint Lucia	-61.083	13.867
Anse Pointe Sable-Man Kote Mangroves	Marine Reserve	Saint Lucia	-60.933	13.733
Bacalar Chico	Marine Reserve	Belize	-87.87447	18.13391
Bodega	Marine Life Reserve	United States	-123.06667	38.31667
Bois D'Orange Mangroves	Marine Reserve	Saint Lucia	-60.95	14.067
Buccoo Reef	Marine Park	Trinidad and Tobago	-60.83385	11.17602
Caesar Point to Mathurin Point reefs	Marine Reserve	Saint Lucia	-60.95	13.717
Cairns	Marine Park	Australia	145.58895	-15.821
Cape D'Aguilar	Marine Reserve	Hong Kong	114.25	22.2
Cas-en-Bas Mangroves	Marine Reserve	Saint Lucia	-60.933	14.083
Choc Bay Artificial Reef	Marine Reserve	Saint Lucia	-80.968	14.034
Choc Bay Mangroves	Marine Reserve	Saint Lucia	-60.969	14.034
Cinque Terre	Marine Reserve	Italy	9.66667	44.16667
Dana Point	Marine Life Reserve	United States	-117.71667	33.46667
Deception Pass	Marine Sanctuary	United States	-122.65136	48.37352
Doheny Beach	Marine Life Reserve	United States	-117.68333	33.46667
Esperance Harbour Mangroves	Marine Reserve	Saint Lucia	-60.917	14.05
Fond D'Or Beach	Marine Reserve	Saint Lucia	-60.883	13.917
Gladden Spit and Silk Cayes	Marine Reserve	Belize	-87.98333	16.5
Glovers Reef	Marine Reserve	Belize	-87.78325	16.82922
Golfo di Portofino	Marine Reserve	Italy	9.25	44.33333
Governor Island	Marine Nature Reserve	Australia	148.31504	-41.87222
Grand Anse Beach and Mangrove	Marine Reserve	Saint Lucia	-60.883	14
Hanauma Bay	Marine Life Conservation District	United States	-157.69635	21.27202
Hervey Bay	Marine Park	Australia	153.00108	-24.94373
Honolua-Mokuleia Bay	Marine Life Conservation District	United States	-156.64369	21.01706
Illa de Tabarca	Marine Nature Reserve	Spain	0.41666	38.16667
Irvine Coast	Marine Life Reserve	United States	-117.81667	33.56667
Kealakakua Bay	Marine Life Conservation District	United States	-155.93064	19.48191
Laguna Beach	Marine Life Reserve	United States	-117.8	33.533
Lapakahi	Marine Life Conservation District	United States	-155.90383	20.17232
Lara-Toxeftra	Marine Reserve	Cyprus	32.25	34.95
Lord Howe Island	Marine Park	Australia	159.18333	-31.66667

Area name	Designate	Country	Longitude	Latitude
International Union for Conservation of Nature and Natural Resources (IUCN) category IV: Habitat/species management areas (continued)				
Louvet Mangroves	Marine Reserve	Saint Lucia	-60.883	13.967
Lundy Island	Marine Nature Reserve	United Kingdom	-4.66667	51.16667
Mackay/Capricorn	Marine Park	Australia	150.98025	-22.38723
Macquarie Island	Marine Park	Australia	161.3338	-55.88075
Managaha Marine Conservation Area	Marine Sanctuary	Northern Mariana Islands	145.697	15.21806
Manele-Hulopoe	Marine Life Conservation District	United States	-156.89492	20.74121
Maria Islet Reef	Marine Reserve	Saint Lucia	-60.93	13.72
Marigot Bay Mangroves	Marine Reserve	Saint Lucia	-61.033	13.967
Marine Conservation Area (Name Unknown) ZAF No.1	Marine Reserve	South Africa	18.49065	-33.8766
Marine Nature Reserve (Name Unknown) ZAF No.1	Marine Nature Reserve	South Africa	28.57471	-32.52112
Marine Nature Reserve (Name Unknown) ZAF No.2	Marine Nature Reserve	South Africa	28.52451	-32.57624
Marquis Mangroves	Marine Reserve	Saint Lucia	-60.9	14.017
Masinloc and Oyon Bay	Marine Reserve	Philippines	119.91687	15.52105
Molokini Shoal	Marine Life Conservation District	United States	-156.5	20.633
Moreton Bay	Marine Park	Australia	153.33915	-27.26687
Moule-a-Chique artificial reef	Marine Reserve	Saint Lucia	-60.947	13.71145
Newport Beach	Marine Life Reserve	United States	-117.85	33.58333
Niguel	Marine Life Reserve	United States	-117.733	33.483
Ninepin Point	Marine Nature Reserve	Australia	147.16733	-43.28413
No Dive Zones (East)	Marine Park	Cayman Islands	-81.18255	19.35015
No Dive Zones (West)	Marine Park	Cayman Islands	-81.22723	19.35274
Old Kona Airport	Marine Life Conservation District	United States	-156.01312	19.64248
Palaui Island	Marine Reserve	Philippines	122.13004	18.54093
Point Firmin	Marine Life Reserve	United States	-118.283	33.7
Port Honduras	Marine Reserve	Belize	-88.60634	16.21469
Praslin Mangroves	Marine Reserve	Saint Lucia	-61.067	13.867
Pupukea	Marine Life Conservation District	United States	-158.03979	21.68512
Reef at Anse de Pitons	Marine Reserve	Saint Lucia	-61.067	13.817
Reef at Malgrétoute	Marine Reserve	Saint Lucia	-61.067	13.833
Reef between Grand Caille and Rachette Point	Marine Reserve	Saint Lucia	-61.067	13.85
Réserve du Larvotto	Marine Reserve	Monaco	7.41667	43.73333
Rodney Bay Artificial Reefs	Marine Reserve	Saint Lucia	-60.967	14.083
Saltwater	Marine Sanctuary	United States	-122.31275	47.36851
San Diego	Marine Life Reserve	United States	-117.25	32.88333
Sapodilla Cayes	Marine Reserve	Belize	-88.2894	16.15002
Savannes Bay Mangrove Area	Marine Reserve	Saint Lucia	-60.917	13.767
Skomer	Marine Nature Reserve	United Kingdom	-5.26838	51.73312
South Laguna Beach	Marine Life Reserve	United States	-117.75	33.51667
South Water Cayes	Marine Reserve	Belize	-88.14643	16.73165
Taklong Island	Marine Reserve	Philippines	122.49888	10.41552
Tinderbox	Marine Nature Reserve	Australia	147.33501	-43.05734
Tobago Cays	Marine Park	Saint Vincent and the Grenadines	-61.35	12.617
Tombant à corail des Spélugues	Marine Reserve	Monaco	7.43333	43.73333
Townsville/Whitsunday	Marine Park	Australia	148.17879	-19.66513
Trinity Inlet/Marlin Coast	Marine Park	Australia	145.65404	-16.70241
Umm al-Maradim	Marine Park	Kuwait	48.65	28.667
Vigie Beach Artificial Reef	Marine Reserve	Saint Lucia	-60.983	14.017
Waikiki	Marine Life Conservation District	United States	-157.8	21.26664
Wailea Bay	Marine Life Conservation District	United States	-155.83291	19.9849
Woongarra	Marine Park	Australia	152.49164	-24.83385

Glossary

Abyssal plain The flat area of an ocean basin between the continental slope and the mid-ocean ridge.

Abyssal zone The ocean between 13,120 and 19,680 feet (4000–6000 m) deep.

Algae Simple plants found as single cells or as seaweeds.

Antarctic circle The line of latitude at 66°33'S marking the northern limit of the Antarctic region.

Aphotic zone The part of the ocean where no surface light can penetrate.

Arctic circle The line of latitude at 66°33'N marking the southern limit of where the Sun does not set in June or rise at December solstices.

Ascidian The sea squirts—a group of invertebrates that produce a larva with a primitive backbone.

Astrolabe An early navigation instrument that was the forerunner of the sextant.

Atoll A coral reef that has formed around a central lagoon.

AUV Autonomous Underwater Vehicle—an unmanned, self-contained submersible.

Backwash The water retreating down the shore after an incoming wave.

Bar A submerged or emerged mound of sand, gravel or shell material built on the ocean floor in shallow water by waves and currents.

Barrier reef A coral reef around islands or along continental coasts, with a deep lagoon between the reef and the coast.

Bathypelagic zone The ocean between 656 and 13,120 feet (200–4000 m) deep.

Bathyscaphe Earliest form of manned submersible.

Bay A recess in the shore or an inlet of a sea between two capes or headlands, not as large as a gulf but larger than a cove.

Beach The region of the shore where loose material, sand, mud or pebble, are deposited between high- and low-water marks.

Benthic zone The seabed.

Berm A horizontal ridge of sand or shingle running parallel to the shore, at the limit of wave action.

Bioluminescence The generation of light by living organisms using the enzyme luciferase.

Bloom The sudden increase in phytoplankton numbers, usually associated with seasonal changes.

Caisson disease The symptoms produced by nitrogen bubbles in body tissues after working in high pressure atmospheres. Also known as the bends.

Cephalopod An advanced group of mollusks that includes the squids, octopuses and cuttlefishes.

Cetaceans The whales and dolphins.

Channel A body of water that connects two larger bodies of water (like the English Channel). A channel is also a part of a river or harbor that is deep enough to let ships sail through.

Chronometer A watch or clock able to maintain its accuracy on long sea voyages.

Coelenterates Gelatinous invertebrates with radial symmetry and sting cells.

Cold seep Cold seawater, rich in methane, hydrogen sulfide and hydrocarbons issuing from the seafloor.

Continental rise Gently sloping base of the continental slope.

Continental shelf The shallow, gently sloping edge of a continental landmass where it meets the sea.

Continental slope The steeply inclined edge of continental plate below the continental shelf.

Crustaceans Invertebrates with jointed limbs and hard chalky shells, such as lobsters, crabs, shrimp and copepods.

Crustal plate A segment of Earth's surface. Continental plates are about 25 miles (40 km) thick and oceanic plates 3.1 miles (5 km) thick.

Current A flow of water in the sea, generated by wind, tidal movements or thermohaline circulation.

Density The mass of a substance for a given volume.

DOC Dissolved Organic Carbon—soluble material given off by organisms that can be absorbed directly as food by others.

Dune An accumulation of wind-blown sand often found above the high-tide mark on sand shores.

Ebb tide Tide period between high and low water. Falling tide.

Echinoderms Exclusively marine invertebrates with five-way symmetry and a water vascular system, including starfishes, sea cucumbers and brittlestars.

Echiurans A group of soft-bodied, non-segmented worms found from the shore down to the bottom of ocean trenches.

Echolocation The use of sound by whales and dolphins to sense objects.

Eddy A circular movement in the water produced by flows around obstructions or by interacting currents.

El Niño The periodic warming of the surface waters in the east Pacific that stops upwelling of nutrients.

Estuary A semi-enclosed area of water where fresh and saltwater mix.

Euphotic zone The upper layers where light is sufficient for photosynthesis.

Fetch The distance over water in which waves are generated by a wind having a rather constant direction and speed.

Flood tide The period of tide between low water and high water. A rising tide.

Fringing reef A coral reef that forms around the shore of an island and gradually extends out to sea.

Gas bladder Gas-filled buoyancy organ in most bony fishes.

Gill Structure used by aquatic animals to exchange dissolved gases and salts between their body fluids and the surrounding water.

Greenhouse effect The warming of the lower layers of the atmosphere caused by the trapping of solar radiation by carbon dioxide and other gases.

Gulf Part of the ocean or sea that is partly surrounded by land, usually on three sides. It is usually larger than a bay.

Gulf Stream The strong western boundary current flowing up the east coast of North America.

Guyot A flat-topped seamount.

Hadal The ocean zone below 19,680 feet (6000 m).

Hermatypic coral Species living in tropical waters that are able to secrete sufficient calcium carbonate to form reefs.

High tide The maximum elevation reached by each rising tide.

Holdfast The multi-branched structure anchoring seaweeds to hard surfaces.

Holoplankton Animals that live out their entire life cycles floating in the water column.

Hurricane An intense tropical cyclone with winds that move counterclockwise around a low-pressure system.

Hydrological cycle The endless cycling of water between land, ocean and atmosphere.

Hydrothermal vent A spring of superheated, mineral-rich water found on some ocean ridges.

Iceberg Floating piece of ice broken off from glacier or ice sheet.

Invertebrate A multicellular animal without a true backbone.

Kelps A group of large, fast-growing brown seaweeds.

Krill A shrimplike crustacean abundant in polar waters that is the principal food of baleen whales.

Lagoon A shallow body of water, such as a pond or lake, usually connected to the sea.

Latitude A measure of north–south location, relative to the Equator at 0°.

Littoral zone Seashore between high and low tide.

Longitude A measure of east–west location relative to the Prime Meridian (0°) that runs through the Greenwich Observatory, London, UK.

Longshore drift The movement of beach material parallel to the coastline by combined wind and wave action.

Lophophore The brushlike feeding organ of sea mats, horseshoe worms and lamp shells.

Low tide The minimum elevation reached by each falling tide.

Magma Molten rock found below Earth's crust that is ejected by volcanoes and emerges at ocean ridges as lava.

Mangrove Flowering shrubs and trees tolerant of saltwater, found on low-lying tropical coasts and estuaries.

Mantle The layer of Earth between the crust and the core.

Mariculture The intensive cultivation of marine organisms in coastal areas in cages or on land in seawater ponds.

Medusa The free-living bell or disklike form of many coelenterates.

Meroplankton The young stages of marine organisms that spend time in the plankton before developing into non-planktonic adults.

Mid-ocean ridge A region of the ocean floor where magma rises to the surface to create new ocean floor on either side of a central rift valley.

Mollusks A group of soft-bodied, non-segmented invertebrates that includes sea snails, bivalves and cephalopods.

Navigation The science of position fixing and course plotting, using astronomical and other observations.

Neap tides Tides with much smaller ranges than spring tides, that occur while the gravitational pulls of the Moon and Sun on the oceans work against each other.

Neritic zone Zone from high tide to the continental shelf break.

Notochord A stiff rod of tissue that becomes part of the true backbone in vertebrate animals.

Ocean One of the five great bodies of seawater defined by continental margins, the Equator and other arbitrary divisions.

Oceanography The scientific study of all aspects of the oceans.

Osmoregulation The regulation of the concentration of body fluids by aquatic animals.

Overfishing The commercial fishing of natural populations so that breeding does not replenish what is removed.

Pack ice Sea ice that forms around the permanent ice sheets of polar regions in winter and which thins and retreats in summer.

Pelagic zone The water column.

Photosynthesis The biological conversion of carbon dioxide and water into sugars using solar energy.

Phytoplankton Single-celled algae and other photosynthetic organisms floating in the surface layers of the oceans.

Pinnipeds The seals, walruses and sea lions.

Plate tectonics The processes by which the plates that form Earth's surface are formed, moved and destroyed.

Polar regions The cold zones between the poles and either the Arctic or Antarctic circles.

Pollutant A harmful substance or heat energy introduced into an ecosystem by human activities.

Polychaete A group of marine segmented worms.

Polynesia A large group of Pacific islands extending from the Hawaiian Islands south to New Zealand and east to Easter Island.

Polyp The sedentary body form of coelenterates, notably corals.

Predator An animal that feeds by capturing and eating other animals.

Primary production The biological conversion of inorganic carbon (carbon dioxide) into living material (organic carbon).

Projection The system used to translate the three-dimensional form of Earth onto a two-dimensional map.

Radar Radio Detection and Ranging—the use of pulsed radio waves to follow moving objects by analyzing changes in reflected radio signals.

Remote sensing The use of airborne or satellite sensors to map Earth's surface in space and time.

Reverse osmosis The use of pressure to force water through a semi-permeable membrane, leaving behind any dissolved salts. Used to obtain freshwater from seawater.

Roaring Forties Areas of ocean either side of the Equator between 40° and 50° N or S latitude, noted for high winds and rough seas.

Rogue wave A single, unusually high wave created by the constructive interference of two or more smaller waves.

ROV A Remotely Operated Vehicle—unmanned submersible controlled and powered from the surface by an umbilical cord.

Salt marsh An area of soft, wet land periodically covered by saltwater, in temperate zones and generally treeless with characteristic salt-tolerant plants such as reeds and samphire.

Scuba Self-Contained Underwater Breathing Apparatus.

Sea A division of an ocean or a large body of saltwater partially enclosed by land. The term is also used for large, usually saline, lakes that lack a natural outlet.

Seamount A steep-sided circular or elliptical projection from the seafloor that is more than ⅔ mile (1 km) in height.

Seasonality The timing of major biological events cued by changes in light intensity and water temperature associated with the seasons in temperate latitudes.

Sediment Fine organic or mineral particles found in the seafloor.

Seismic survey The use of high-intensity sound waves to examine deep geological structures.

Sextant A navigational instrument used to measure angles between the Moon, Sun, stars and objects such as the horizon.

Side-scan sonar High-resolution sound-imaging of the seabed.

Sipunculans Peanut worms—a group of non-segmented worms thought to be related to the polychaetes.

Shelf sea The shallow but often highly productive seas over continental shelves.

Sonar Sound Navigation and Ranging—the detection of objects in or on water using pulsed beams of sound waves and their reflected echoes.

Spring tide A tide that occurs at or near the time of a new or full Moon with a large tidal rise and fall.

Sponges Invertebrates that consist of complex aggregations of cells, bound together by protein fibers and mineral spicules.

Strait Narrow channel of water that connects two larger bodies of water, and thus lies between two landmasses.

Subduction zone An area where one crustal plate is forced under another plate, giving rise to volcanic activity and earthquakes.

Submersible A small underwater vehicle designed for deep sea research and other tasks.

Symbiosis The close beneficial feeding relationship between two species.

Thermohaline circulation Water movement caused by differences in density produced by salinity and/or temperature changes.

Tide The regular rising and falling of the sea that results from gravitational attraction of the Moon, the Sun and other astronomical bodies acting upon the rotating Earth.

Trade winds Steady winds blowing east to west toward the Equator to replace hot air rising from the equatorial region.

Trench A narrow, deep depression in the ocean floor, often associated with the subduction of an oceanic plate at a continental margin.

Trophic web The complex feeding relationships between plants and animals in a habitat.

Tropics The zone between the Tropic of Cancer (23°27′N) and the Tropic of Capricorn (23°27′S) approximating to the area of ocean where water temperatures remain above 69°F (20°C).

Tsunami A huge wave created by earthquake or volcanic explosion. Mistakenly called a tidal wave.

Typhoon A hurricane in the western Pacific Ocean or China seas.

Upwelling The rising of deep, cold-nutrient-laden waters into the surface layers, close to continental coasts.

Wave The disturbance in water caused by the movement of energy through the water.

Zooanthellae Single-celled photosynthetic organisms that live in coral tissues in a symbiotic relationship.

Zooplankton Small animals that spend all or part of their life cycles floating in the surfaces layers of the ocean.

Index

Credits

PHOTOGRAPHS

t=top; l=left; r=right; tl=top left; tc=top center; tr=top right; cl=center left; c=center; cr=center right; b=bottom; bl=bottom left; bc=bottom center; br=bottom right

AAA = The Ancient Art & Architecture Collection Ltd.; AAP = Australian Associated Press; AFP = Agence France-Presse; APL/Corbis = Australian Picture Library/Corbis; APL/MP = Australian Picture Library/Minden Pictures; AUS = Auscape International; COML = Census of Marine Life; COR = Corel Corp.; GI = Getty Images; IQ3D = Image Quest 3-D; LT= Lochman Transparencies; N_EO = NASA/Earth Observatory; N_ES = NASA/Earth from Space; N_V = NASA/Visible Earth; NASA = National Aeronautics and Space Administration; NOAA = National Oceanic and Atmospheric Administration; NPL=Nature Picture Library; NV = naturalvisions.co.uk; NWU = Norbert Wu Productions; PL = photolibrary.com; SP = Seapics.com; UWP = UWPhoto

Front Cover tl PL; tc, tr, bl, br GI
Spine Peter Bull Art Studio
Back Cover Moonrunner Design

1c PL **2**c APL/Corbis **4**c GI **6–7**c PL **8–9**c APL/Corbis **10**c APL/Corbis cl N_ES cr GI **11**c, cr GI cl IQ3D/Y Kito **12**c APL/Corbis **14–15**c N_ES **16**c N_V cl APL/Corbis cr PL **17**c N_V/Jacques Descloitres/MODIS Rapid Response Team/GSFC cl N_EO/Serge Andrefouet/University of South Florida **18**c N_EO/GSFC **23**bl PL t GI **24**tr NASA/Our Earth as Art **25**b N_EO/Serge Andrefouet/University of South Florida cr APL/Corbis **34**bl APL/Corbis **35**tr APL/Corbis **36**bl APL/Corbis **37**tr PL **38**cl APL/Corbis tr GI **39**t GI **40**bl APL/Corbis **42**t APL/MP **43**b APL/Corbis t SeaWiFS Project/NASA/GSFC/ ORBIMAGE **44**tr Jerome Cuny **45**r APL/Corbis **46**br APL/Corbis tr GI **47**c APL/Corbis **48**bl P br AAP/AFP tr APL/Corbis **49**bc, c, tc PL, r NOAA **50**bl, tr APL/Corbis tc PL **52**br, cl, tc, tr APL/Corbis **54**cr N_V **55**br AFP tr PL **56**bl, c N_V/Jacques Descloitres/MODIS Rapid Response Team/GSFC **57**br N_V/Jacques Descloitres/MODIS Rapid Response Team/GSFC **58**c APL/Corbis **59**br APL/Corbis tr PL **60**cl PL cr LT/ Eva Boogaard **61**br, tr PL **62**b, tr APL/Corbis **63**bc PL **64–65**c APL/ Corbis **66**c, cl APL/Corbis **67**c, cl APL/Corbis **68**bc, br APL/Corbis bl The Art Archive/Musée du Louvre Paris/Dagli Orti **69**bc, bl, br APL/Corbis **70**bl, cr APL/Corbis **71**b, tr APL/Corbis **72**bl, br APL/ Corbis tl AAA tr 2004 Photo SCALA, Florence **73**cr GI **74**bl APL/ Corbis br By permission of the National Library of Australia **75**bl APL/Corbis **76**tr APL/Corbis **77**b APL/Corbis **78**b AAP/AFP **79**b AAP c AAP/90 t APL/Corbis **80**bl, cr APL/Corbis **82**bc GI bl, br APL/Corbis **83**bc, bl APL/Corbis br PL **84**bc The Alfred Wegener Institute bl Australian Institute of Marine Science br Woods Hole Oceanographic Institution **85**bc Southampton Oceanography Centre bl Harbor Branch Oceanographic Institute tr Japan Marine Science and Technology Center **86**br, cl, tr PL **87**c PL t N_V **88**bl NOAA Office of Marine and Aircraft Operations cr APL/Corbis **89**br PL **90**tr APL/Corbis **92**bc, bl PL br APL/Corbis **93**bc, bl PL br APL/Corbis **94**b, bl, tr APL/Corbis **95**tr APL/Corbis **98**bl APL/Corbis cr PL **99**br APL/Corbis tl, tr J Copley **100**bl PL **102**bl PL **103**c GI **104–105**c GI **106**c SP/Saul Gonor cl PL cr GI **107**c, cl APL/Corbis **110**bl PL r IQ3D/Peter Parks **111**br IQ3D/Peter Parks tr SP/Peter Parks **112**bl, cr APL/Corbis **113**br PL tr APL/ Corbis **114**cr APL/Corbis **115**br APL/Corbis cr IQ3D/Carlos Villoch tr NWU **116**b GI **117**cr APL/Corbis tl PL **118**bl IQ3D/Chris Parks br NV/Norman T Nicoll tr IQ3D/Peter Parks **119**r PL **120**bl APL/MP cr PL **121**br APL/Corbis tr IQ3D/Musa Ushioda **122**bl IQ3D/Peter Parks cl SP/Marc Chamberlain tr UWP/Erling Svensen **123**br IQ3D/Peter Parks **124**bl COR br, t APL/Corbis **125**bl LT br PL **126**bc, bl APL/MP br SP/C & M Hochleitner/V&W t SP/Doug Perrine **127**bl, br COR cr APL/MP **128**cr APL/Corbis **129**bl SP/ James D Watt br, tr PL **130**br PL tr APL/MP **131**br NV/Heather Angel tr APL/MP **132**bl IQ3D/Peter Batson br AUS/Yves Lanceau t NPL/Reijo Juurinen/Naturbild **133**br Photo Researchers Inc./ Berthoule-Scott cr APL/Corbis **134**bl GI br, tr APL/MP **135**br UWP/ Erling Svensen **136**IQ3D/Carlos Villoch br IQ3D/Masa Ushioda t APL/Corbis **137**bl, cr APL/Corbis **138**br COR SP/Rudie Keiter **139**br, tr APL/Corbis **140**bl SP/Garry Bell br GI tr IQ3D/Scott Tuason **141**cr GI

tr APL/MP **142**r APL/Corbis **143**br, cr NPL/Jurgen Freund **144**bl GI tr APL/MP **146**bc, bl APL/Corbis br APL/MP **147**bc, bl, br APL/Corbis **148**b APL/MP **149**cl SP/Saul Gonor cr SP/ Amos Nachoum tr APL/MP **150**bl, br APL/Corbis tr APL/MP **151**bc, bl APL/MP br APL/Corbis **154–155**c IQ3D/Y Kito **156**c IQ3D/Peter Herring cl IQ3D/Justin Marshall cr Dave Wrobel **157**c AUS/Norbert Wu c P Batson/ExploreTheAbyss.Com **158**bl AAP/ AFP **160**bl APL/Corbis **161**br, t PL **162**tr R Lampitt **163**c SP/Bob Cranston **164**bl NPL/David Shale br IQ3D/Peter Batson tr IQ3D/Peter Herring **165**b IQ3D/Peter Parks tr IQ3D/Peter Herring **166**bl NPL/David Shale c IQ3D/Peter Herring tr SP/Bob Cranston **167**bl IQ3D/Peter Parks br NPL/Conrad Maufe **168**br APL/MP cl IQ3D/Justin Marshall cr NPL/David Shale **169**br NPL/Florian Graner tr OAR/ National Undersea Research Program/Harbor Branch Oceanographic Institution **170**cl IQ3D/Peter Batson cr AUS/ Norbert Wu **171**bl, cr AUS/Paulo De Oliveira tl IQ3D/Peter Herring **172**bl P Batson/ExploreTheAbyss.Com cr NPL/David Shale **173**br NOAA/National Undersearch Research Program (NURP) Collection tr IQ3D/Peter Herring **174**bl Dave Wrobel tr IQ3D/Peter Herring **175**bl, cr IQ3D/Peter Herring **176**b APL/MP tr NOAA/OAR/National Undersea Research Program (NURP)/ University of South Carolina **177**br Dave Wrobel **178**bl APL/ Corbis tr NPL/David Shale **179**br, tr IQ3D/Peter Herring **180**bl NWU br IQ3D/Peter Herring tr IQ3D/Justin Marshall **181**bl IQ3D/ Justin Marshall tr APL/Corbis **182**bl NWU **183**b Picture taken with the ROV Victor 6000 of Ifremer, at 2500 meter depth, copyright Ifremer/biozaire2-2001 tr APL/Corbis **184**bl IQ3D/Y Kito **185**br NWU cl IQ3D/Chris Parks tr IQ3D/Peter Herring **186**b, tr IQ3D/Peter Herring **187**br IQ3D/Peter Herring t NWU **188**bl AUS/Clive Bromhall cr NWU **189**b NPL/David Shale t APL/Corbis **190**bl P Batson/ExploreTheAbyss.Com cr NOAA/OAR/National Undersea Research Program (NURP) **191**br Census of Marine Life/Ian MacDonald/Texas A&M University Corpus Christi tr PL **192**b NOAA/OAR/National Undersea Research Program (NURP)/ Texas A&M University tr NOAA/OAR/National Undersea Research Program (NURP)/College of William & Mary **193**b Dave Wrobel tr NOAA/OAR/National Undersea Research Program (NURP)/Texas A&M University **194–195**c GI **196**c SP/Mark Conlin cl APL/Corbis cr SP/Phillipa Cola **197**c, cl PL **198**bc APL/MP bl UWP br APL/Corbis **199**bc, br PL bl UWP **200**bl, br APL/Corbis t GI **202**bl PL br, tr APL/Corbis **203**cr, tl APL/Corbis **204**bl PL br AFP t APL/Corbis **205**bc GI **206**bl PL cr APL/Corbis **207**cr GI tl APL/MP **208**bl N_ES c NASA/GSFC/METI/ERSDAC/JAROS & US/Japan ASTER Science Team **210**bl NV cr APL/Corbis **211**br, tl PL **212**b APL/Corbis tr NASA/Landsat 7 **213**br APL/Corbis cl GI **214**bl PL br APL/Corbis t APL/MP **215**br APL/Corbis **216**bc GI bl IQ3D/Valdamir Butterworth cl APL/Corbis **217**br APL/MP cr IQ3D/Jim Greenfiels tl SP/Phillipa Cola **218**bl APL/MP br, tr APL/ Corbis **219**br APL/MP tr SP/Mark Conlin **220**bl, tr APL/Corbis **221**cr PL tr APL/Corbis **222**bl PL c GI **224**bc, br, cl APL/Corbis tr GI **225**br APL/Corbis **226**b APL/Corbis tr NPL/David Kjaer **227**cl APL/Corbis r NPL/Richard Du Toit **228**bl, br, t APL/Corbis **230**bl, br, cr APL/Corbis **231**cl APL/Corbis tr LT **232**bl NWU br APL/Corbis tr PL **233**br, tr APL/Corbis **234–235**c GI **236**c, cl, cr APL/Corbis **237**cl GI **238**bl, br PL tr APL/Corbis **239**br APL/Corbis **240**b, tr APL/Corbis **241**c APL/Corbis **242**bl AFP tr APL/Corbis **243**bl NPL/Ashok Jain cr APL/Corbis **244**bl, br, tr APL/Corbis **246**bl PL cr APL/Corbis **247**tr AFP **248**bl APL/Corbis **249**bl, cr APL/Corbis tl GI **250**br, cl, tr APL/Corbis **252**bl, c, cr, tr APL/Corbis **254**cl APL/Corbis tr Jeremy Sutton-Hibbert **255**br APL/Corbis **256**bl, cr APL/Corbis **258**bl, br PL tr APL/Corbis **259**br APL/Corbis **260**bl APL/Corbis tr PL **261**bl PL **262**bl APL/Corbis tr PL **263**bl New Forest District Council **264**bl, tr PL **265**t GI **266**b APL/Corbis tr AAP/AFP **267**t AAP/AFP **268**b APL/Corbis tr PL **269**tl, tr PL **270**bl, tr APL/Corbis br IQ3D/Chris Parks tc APL/MP **271**br PL **272**tr, c PL **273**br PL **274**bl, br, tr APL/Corbis cl NASA/TOMS **276**bl SP/Doug Perrine br, tr APL/Corbis **277**tc GI tr APL/Corbis **278**bl UWP/Erling Svensen cr IQ3D/James D Watt **280–281**c 2004 Photo SCALA, Florence

ILLUSTRATIONS

Peter Bull Art Studio: 18bl, 19br, 38br, 39b, 50br, 53br, 54bl, 89tc, 90b, 108c, 112cl, 114b, 117bl, 119br, 120c, 123tr, 125cr, 128cl, 130cl, 144br, 145br, 152cl, 158r, 160cr, 162b, 179cl, 181tl, 182tr, 184tr, 187cl, 191cl, 201br, 205cr, 207br, 209r, 211cl, 220bl, 223cr, 231br, 248tr, 257cr, 260br, 263r, 264br, 265br

Andrew Davies Creative Communication: 19tr, 24bl, 28c, 30c, 32c, 41b, 49r (adaptation), 147cl, 254b, 261b, 275br

Chris Forsey: 22tr

Jon Gittoes: 229r

Map Illustrations & Andrew Davies Creative Communication: 22b, 26-27, 28cl, r, 30cl, r, 32cl, r, 34cl, c, 36cl, c, 40br, 42b, 44bl, 57tr, 58cl, 74t, 75r, 76br, 84tr, 142bl, 146tr, 198tr, 238cl, 256cl

Moonrunner Design: 40tr, 66cr, 96c, 100cr, 102tr

Wildlife Art Ltd./Richard Bonson: 20-21

Wildlife Art Ltd./Tom Connell: 45bl, 46bl, 87b

Wildlife Art Ltd./Mick Posen: 252br

Captions

page 1 Brightly colored reef fishes dart among the bleached branches of stag horn coral on a tropical reef.

page 2–3 Sunlight filters through the canopy of fronds in a giant kelp forest in the waters off the coast of California.

page 4–5 As waves approach the shore they rise up, curling into a tube before crashing down on to the beach.

page 6–7 A spider crab picks its way through the branches of a red coral in search of scraps left by the coral polyps.

page 8–9 For a few weeks each year, fur seals come ashore in vast numbers on a favored secluded beach to mate and raise their pups.

page 12–13 Pacific Island men paddle an outrigger canoe. Pacific Islanders were one of the first peoples to explore the oceans—searching for riches and colonizing new islands.

page 280–281 This fresco of a Minoan flotilla from around 1650 BC shows the great importance of the sea to ancient Mediterranean cultures.